T0282630

CAMBRIDGE LIBRARY COLLECTION

Books of enduring scholarly value

Earth Sciences

In the nineteenth century, geology emerged as a distinct academic discipline. It pointed the way towards the theory of evolution, as scientists including Gideon Mantell, Adam Sedgwick, Charles Lyell and Roderick Murchison began to use the evidence of minerals, rock formations and fossils to demonstrate that the earth was older by millions of years than the conventional, Bible-based wisdom had supposed. They argued convincingly that the climate, flora and fauna of the distant past could be deduced from geological evidence. Volcanic activity, the formation of mountains, and the action of glaciers and rivers, tides and ocean currents also became better understood. This series includes landmark publications by pioneers of the modern earth sciences, who advanced the scientific understanding of our planet and the processes by which it is constantly re-shaped.

An Elementary Introduction to the Knowledge of Mineralogy

William Phillips (1773–1828) was a printer and geologist who became a Fellow of the Royal Society in 1827. A founder of the London Askesian Society, he was also an active member of the British Mineralogical Society. In 1807 he and twelve others founded the Geological Society of London, and he was described by the Society's historian as 'the most distinguished, as a geologist, of the original founders'. His pioneering 1818 digest of British geology, *Outlines of the Geology of England and Wales*, was the most ambitious and influential work of its kind. Phillips gave free lectures to young people in his village in 1814, and these were published the following year. This work followed in 1816, and both went on to become standard textbooks. Aimed at students, it collects observations of a wide range of minerals' characteristics and occurrence, incorporating crystallographic work using the new reflecting goniometer.

Cambridge University Press has long been a pioneer in the reissuing of out-of-print titles from its own backlist, producing digital reprints of books that are still sought after by scholars and students but could not be reprinted economically using traditional technology. The Cambridge Library Collection extends this activity to a wider range of books which are still of importance to researchers and professionals, either for the source material they contain, or as landmarks in the history of their academic discipline.

Drawing from the world-renowned collections in the Cambridge University Library and other partner libraries, and guided by the advice of experts in each subject area, Cambridge University Press is using state-of-the-art scanning machines in its own Printing House to capture the content of each book selected for inclusion. The files are processed to give a consistently clear, crisp image, and the books finished to the high quality standard for which the Press is recognised around the world. The latest print-on-demand technology ensures that the books will remain available indefinitely, and that orders for single or multiple copies can quickly be supplied.

The Cambridge Library Collection brings back to life books of enduring scholarly value (including out-of-copyright works originally issued by other publishers) across a wide range of disciplines in the humanities and social sciences and in science and technology.

An Elementary
Introduction
to the Knowledge
of Mineralogy

WILLIAM PHILLIPS

CAMBRIDGE
UNIVERSITY PRESS

CAMBRIDGE UNIVERSITY PRESS

Cambridge, New York, Melbourne, Madrid, Cape Town,
Singapore, São Paolo, Delhi, Mexico City

Published in the United States of America by Cambridge University Press, New York

www.cambridge.org
Information on this title: www.cambridge.org/9781108049382

© in this compilation Cambridge University Press 2012

This edition first published 1816
This digitally printed version 2012

ISBN 978-1-108-04938-2 Paperback

AN

ELEMENTARY INTRODUCTION

TO THE KNOWLEDGE OF

MINERALOGY:

INCLUDING

SOME ACCOUNT OF MINERAL ELEMENTS AND CONSTITUENTS ;

EXPLANATIONS OF TERMS IN COMMON USE;

BRIEF ACCOUNTS OF MINERALS, AND OF THE PLACES AND
CIRCUMSTANCES IN WHICH THEY ARE FOUND.

DESIGNED FOR THE USE OF THE STUDENT.

By WILLIAM PHILLIPS,

MEMBER OF THE GEOLOGICAL SOCIETY.

Nullum est sine nomine saxum.   Lucan.

LONDON:

PRINTED, AND SOLD BY WILLAM PHILLIPS,
GEORGE YARD, LOMBARD STREET.

1816.

PREFACE.

A considerable edition of the little volume en-
titled an ' Outline of Mineralogy and Geology' having been nearly, and very unexpectedly, exhausted
in the short space of a few months, it became a
subject of consideration whether it could be so enlarged as to render a second edition more interesting and valuable, without greatly increasing the
size and price of the book.   It occurred to me that
it might be possible further to illustrate the subjects
on which it treats, by introducing some account of
the more important minerals, with general observations upon such as should be omitted.   But, when
considerable progress had been made, it assumed
such a patch-work character, that I resolved to re-
publish the ' Outline' with such alterations only
as might seem essential to be made ;   more especially as partial descriptions would completely

alter its character, without fully answering the purpose of their introduction. Its object will there- fore be, as before, rather to awaken inquiry, than to satisfy it.

The attempt to improve it, however, served to convince me that if descriptions of individual mi- nerals, together with some account of the places and circumstances in which they are commonly found, were collected with tolerable fidelity from the best authorities, and comprized in a small volume, it would prove instructive to the young mineralogist ; more especially if divested of technical and scientific terms as much as the nature of the subject will allow. This feeling was an incitement to undertake the labour of selecting, which at best is but an humble occupation.

It next became an object to determinate the order in which these descriptions should be placed : and when it is considered that any one of the se- veral arrangements that have already been promul- gated, might have been chosen, I can scarcely hope to escape censure for having adopted one that in some respects differs from them all ; my apology is, that not one of them was adapted to my purpose.

In the introduction to Aikin's ' Manual of Mi- neralogy,' an attentive perusal of which I wish again to recommend to the student, there are some excellent remarks on the prevailing arrangements of minerals. From all that has hitherto been done, it may be argued, that the very nature of the sub- stances comprehended in those arrangements, forbids the construction of any one against which many well-founded objections cannot be raised ; and wherein there shall not be much that is arbitrary, and consequently, dependent on some particular views, or some favorite theory. Hitherto, no natural classification of minerals has been discovered : either this most desirable object cannot be attained, or the science is not yet sufficiently understood to allow of its accomplishment.

It cannot, however, be denied, on the one hand, that the science of mineralogy is greatly dependent on that of chemistry ; nor, on the other, that its acquirement should be regarded as preliminary to that of geology. It therefore seemed indisputable, that if it were possible to exhibit the science in such a point of view, as that its dependence on the one, and its intimate connexion with the other, should become apparent, the result would be advantageous to the student.

With these objects principally in view, peculiar attention has not always been given to the enumeration of all the nicer characteristics of each mineral, nor to the maintaining of one exact order of description. This has been done by Aikin in his 'Manual of Mineralogy,' with all the fidelity which a precise and scientific detail of these characters requires. It has rather been my intention to give in familiar language, the more important mineralogical and geological characters of each ; so as to enable the student, by such acquaintance as he may thus familiarly gain with the objects of his to study, to consult with advantage the more scientific works of abler mineralogists. Consistently with this intention, explanations of about one hundred terms commonly used in mineralogical description, are given at the end of the Introduction ; which includes an enumeration of mineral elements and constituents, together with a brief view of their chemical characters, and remarks tending to shew their mineralogical and geological importance. The order in which the individual minerals have been described, and which is exhibited in the Table of Contents, was governed by an attention to the same objects.

Thus siliceous minerals are first described, because it is estimated that silex forms the largest proportion of the oldest and most abundant primitive rocks : and all earthy minerals of which silex is the largest ingredient, are arranged under that head ; beginning, chemically, with silex in its purest form, and proceeding to such as consist of that and another earth, as silex and alumine, then to those consisting of silex and lime, &c. and afterwards to such minerals as are chiefly constituted of three or more earths, terminating with the most compound ; and regarding the iron, manganese, &c. involved in many of them, only as accessary constituents. The other earthy minerals are proceeded with in like manner ; arbitrarily selecting such as contain the rare earth glucine, and placing them under that head, except that the Gadolinite, which also contains the still more rare earth, Yttria, is placed under the latter. In regard to metalliferous minerals, the rules I had prescribed for the order of description could not always be adhered to without

involving some absurdity; for instance, in the ore
called White Silver, that metal is an ingredient, according to one analysis by Klaproth, in the proportion only of about 2 per cent.; but it would
have been altogether ridiculous to have placed a
substance bearing the name of White Silver among
the ores of Lead, to which, according to the proportions of its ingredients, it properly belongs.

In order to avoid too greatly the appearance of a
scientific work, every mineral has been described
only under the name or names by which it is commonly distinguished in our own country. The sy-
nonyms may be found in Aikin's ' Manual,' and
still more at large in the useful ' Mineralogical
' Nomenclature' of Allan.

This compilation, for it includes but little that
is new, has been selected from the works of the
most experienced mineralogists; chiefly from those
of Haüy, Brongniart, Jameson, the Chemical and
Mineralogical Dictionary of Aikins, and the Manual of Aikin; but, considering the purpose of
the book, it seemed unnecessary to acknowledge
the numerous quotations from those and other
works in a more particular manner, by repeated
references to their pages.

It is common with the beginner to ask for some
means by which he may be enabled at once to recog-
nize any mineral that may present itself to his notice.
To this inquiry, it may be replied, that, without
the aid of study and experience, no means sufficiently
precise can be hoped for in a science which is
without a natural arrangement; and which there-

fore is divested of the certainties belonging to the study of the animal and vegetable kingdoms. A studious comparison of their characters, with the descriptions published in the works of mineralogists, might possibly enable the student to accomplish this desirable object. This method is undoubtedly tedious ; and in most cases, the best rules that have been laid down, presuppose certain previous attainments ; but the labour would be materially lessened, if the individual specimens were well characterized, and properly designated. The most effectual and advantageous method of acquiring a competent knowledge of minerals is undoubtedly that of personal instruction. The superiority which France and Germany have acquired in mineralogical science is, doubtless, in a great measure to be attributed to the facility of obtaining instruction, both public and private ; of which there was an almost total deficiency in this country, until very lately. Each of our Universities has now its professor, and private instruction begins to be attainable. The metropolis and its neighbourhood are not without advantages in this respect. Lectures are given at the Royal and Surry Institutions. The time and attention of Mrs. Lowry, of Titchfield Street, whose fine collection of minerals, models, and instruments used in the mineralogical and geological researches, cannot fail, under her instruction, of being advantageous to her pupils, are occasionally given to this object: and T. Webster, of Buckingham Place, Fitzroy Square, who is draughtsman to the Geological Society, and has the immediate care of its valuable collection, and whose acquirements may thence be estimated, also dedicates a part of his time to in- struction in the sciences of mineralogy and geology, as well as to the teaching of drawing; a knowledge of which is intimately connected with those sci- ences, and in the instruction of which he has adopted the most expeditious and advantageous methods he can devise.

Instruction in crystallography is also attainable. N. J. Larkin, of Gee Street, Somer's Town, who is a teacher of the mathematics, is in the habit of teaching their application to the theory of crystal- lization of Haüy. A perfect knowledge of this most beautiful theory can only be attained by a correct acquirement of the mathematical principles on which it is founded; nevertheless, the theory is also taught mechanically by N. J. Larkin, in a few lessons, by the assistance of models. These models, cut in box-wood, may be had of Bate in the Poul- try, and Mawe in the Strand, at one guinea each, as well as complete sets of models of all the crystals described by Haüy in his Treatise on Mineralogy, from eight pounds to sixteen pounds the set, ac- cording to the kind of wood of which they are made: they are cut by N. J. Larkin with great accuracy and beauty.

In the descriptions of individual minerals in- cluded in this volume, it was my wish to have given a somewhat detailed account of their crystalline forms. This I found to be impossible without in- creasing the size of the book considerably. As, however, I conceive that it would materially tend to facilitate the progress of the young mineralogist, it is my intention at some future time to publish a view of the theory of crystallization, unless it shall be accomplished by some abler hand. This view will not be illustrated by the application of its mathematical principles, and will therefore be only mechanical; but it will necessarily be accompanied by numerous figures, illustrative of the theory, and of the transitions of crystalline forms.

It is probable that some who may look into this volume, may judge that if the descriptions had been more at length, more precise, and more technically scientific, they would have been more valuable, and consequently of more general interest. Such as may be induced to pass this judgment upon it, are entreated to advert to the main purpose of the publication. But, the simplicity of the design, and in all probability the manner in which that design is executed, will deter the scientific from perusing a work which is manifestly intended only for the beginner—only as a first step for the student—and which in reality has little claim to the notice of the mineralogist.

W. P.

London, March, 1816.

INTRODUCTION.

———

THE investigation of the structure of the earth belongs to the science of Geology. It may however be interesting to take a rapid survey of the present state of our knowledge respecting it, were it only for the sake of shewing its intimate connexion with mineralogical pursuits.

In speaking of the earth and of our knowledge of its structure, it is essential that the limited extent of that knowledge should always be had in remembrance. We are acquainted with it, only to a very inconsiderable depth; and when it is recollected that, in proportion to the bulk of the earth, its highest mountains are to be considered merely as the unimportant inequalities of its surface, and that our acquaintance does not extend in depth, more than one-fourth of the elevation of these mountains above its general level, we shall surely estimate our knowledge of the earth to be extremely superficial; that it extends only to its crust.

The term ' Crust of the Earth' therefore relates only to the comparative extent of our knowledge

a

beneath its surface. It is not used with the intention of conveying an opinion that the earth consists only of this crust, or that its center is hollow ;
for of this we know nothing. The term may not
be philosophical, but it is convenient.

The structure of the crust of the earth is most
readily studied in mountains, because their masses
are obvious ; and also because, as they are the chief
depositories of metalliferous ores, the operations of
the miner tend greatly to facilitate their study.
Mountains are composed of masses which have no
particular or discernible shape : or, as is more
commonly the case, of strata or beds, either horizontal or oblique, sometimes nearly vertical.

In these masses and beds different structures
have been observed. Some of them are crystalline ; that is to say, are composed of crystals deposited in a confused manner, as in granite, or of
crystals imbedded in some other substance, as in
porphyry. These crystalline rocks contain no
organic remains ; and, as they are always found
beneath, never above, those which do contain
them, they are considered to have been of earlier
formation, and therefore have been termed *primitive rocks.*

Other mountain rocks have no appearance of crystallization ; but, on the contrary, seem rather to
have been formed by the mere falling down, or set-
tlement, of the substances of which they are composed, from the solution which contained them.
These are always found above, never beneath, the
crystalline rocks ; and often contain a vast abun-

dance of organic remains, both animal and vegetable.
The more ancient of these, or such as contain the
remains of animals of which the genera and species are extinct, are called *Transition rocks:* the
more recent, or such as contain the remains of
animals in some degree, or perfectly, resembling
those now inhabiting our oceans, are called *Flœtz*
or *Flat* rocks, because their position is considerably, or perfectly horizontal: the former have received the name of Transition, as connecting the
primitive with the flœtz rocks. By many mineralogists the transition and the flœtz are classed together under the name of *secondary rocks.*

Primitive and secondary rocks have suffered considerable change and ruin from causes which it is
not our present object to notice; and their disintegrated portions, having been formed anew, now
constitute that peculiar description of deposite
which is termed *alluvial,* and which therefore consists of the *debris* of other rocks. Such are clays,
gravel, sand, &c. and these often contain the remains of land and amphibious animals, and of fish:
they are found above the preceding, sometimes
resting immediately upon primitive rocks.

But there is still another and a very different kind
of rock, abundantly found in certain countries,
which may in great measure be considered, like
the preceding, as resulting from the ruin of rocks,
but from an opposite cause, or by an agent directly
the reverse, viz. by fire; constituting those known
by the name of *volcanic rocks:* many of these
strongly bear the marks of heat, and even of fu-

sion; some, on the contrary, offer no evidence of
their having been subjected to heat.

Lofty mountains composed of primitive rocks
usually present rugged and uneven summits, and
steep acclivities on the sides, as though they had
suffered by convulsion. Such as are wholly or externally composed of secondary beds or strata, are
less rugged on the summits and sides ; their summits are flattish, or somewhat rounded, and their
sides present acclivities more easily accessible ; and
are still more so when covered by alluvial matter,
which serves to fill up their roughnesses and hollows, and often presents nearly a plane surface

Both primitive and secondary mountains, more
particularly the former, are traversed in various
directions by fissures, of different dimensions.
These fissures are not often empty, but are mostly
filled with stony or metalliferous substances, accompanied by vast quantities of water; but not
often by portions of the rocks they traverse. These
fissures are termed *Mineral Veins*: of whatever
substance or substances, the body of a vein may
be composed, its sides are commonly very determinate, and are by the miner called the walls of
the vein.

From these veins, a large proportion of all the
minerals which are found in the cabinet of the
mineralogist, are extracted ; indeed almost all such
as, from their rarity, brilliancy, or peculiarity of
form and combination, possess the greatest attrac-
tion for the mere collector: but these, though in

these respects they may be the most curious, are by no means the most important.

Mineralogy is a science of so great interest, that it would be much to be regretted were its real objects and tendency misunderstood, or suffered to degenerate into an avidity merely for the collecting of what is brilliant or rare. It is capable of affording larger and more useful attainment than the possession of an unique. To the attainment of the science of geology, that of mineralogy is essentially requisite.

The study of mineralogy therefore does not include only a knowledge of the more rare and curious substances; there is nothing in the mineral kingdom too elevated or too low for the attention of the mineralogist, from the substances composing the summits of the loftiest mountains, to the sand or gravel on which he treads. It is true that the aggregated masses of compound rocks are not arranged in a mineralogical collection; but it must be remembered that each of the substances of which such aggregated masses are constituted, are all comprehended in a mineralogical arrangement, and therefore find their places in the cabinet. Granite, it is true, is not to be found there; but its components, quartz, felspar, and mica, are met with in every one.

Thus, then, by the study of what, in opposition to the term aggregated rocks, may be termed simple minerals, the mineralogist becomes enabled to detect the substance with which he holds acquaintance by itself, when aggregated with others in a mass; and thus he becomes qualified for the more difficult, and more important study of the science of geology; which embraces a knowledge of the nature and respective positions of the masses and beds composing mountains; and indeed of country of every description, whether mountainous or otherwise.

It is not, therefore, or at least it ought not to be, the sole object of the mineralogist, to be able to distinguish the several genera and species of mineral substances; nor should his attention be confined to the mere task of recognizing at first sight any mineral that may present itself, or of being capable of at once assigning it a proper place in his cabinet. He should hold a more enlarged acquaintance with minerals, and with the circumstances attending them, in what may be termed, their native places; he should know something of the positions they respectively bear towards each other in those places; he should become acquainted with their relative ages, deduced from the nature of the rocks in which they are found; their comparative scarcity or abundance; their combinations; the *countries* in which they occur; and their characters, both internal and external.

This knowledge, it may be repeated, is the first and requisite step to the science of geology: not that it is essential to this science that every mineral should be accurately known: some are of comparatively little importance in a geological point of view, from their extreme scarcity; but it is essential to become acquainted with simple mine a in the general, because of some of them, many of the vast masses of the earth are composed.

Minerals which are found only in primitive rocks, are said to belong to *primitive countries;* by which name are designated such tracts as are chiefly composed of primitive rocks. The substance in or on which a mineral is found, is called its *gangue* or *matrix ;* when in its natural place or position, a mineral is said to be *in situ;* when this place and position are known, we are acquainted with its *habitat.*

In conformity with the object of this work, as explained in the preface, we must, before entering upon a description of individual minerals, take a view of the number, as well as of the nature of the elementary bodies, of which they are constituted. In this, I shall aim at brevity.

The whole number of mineral elements are commonly included in the list of 9 earths, 2 alkalies, 27 metals, and the two bases of combustible bodies, carbon and sulphur ; but there are still other substances, both simple and compound, which having been detected by analysis, as entering into the composition of certain of the minerals about to be described ; it seems essential in an elementary view of the science that these constituents should have a due consideration, whether they be regarded essentially as mineral elements, or only as accessaries.

These substances consist of certain acids, together with water, hydrogen, and oxygen.

The acids are 13 in number, and are compound
substances; generally, though not without excep-
tion, consisting of oxygen, united in different proportions with certain bases.

The base of the *Molybdic* acid is *Molybdena.*
.... *Arsenic* ...... *Arsenic*
.... *Chromic* ...... *Chrome*
.... *Tungstic* ...... *Tungsten*
.... *Carbonic* ...... *Carbon*
.... *Sulphuric* ...... *Sulphur*
.... *Phosphoric* ...... *Phosphorus*
.... *Fluoric* ...... *Fluorine*
.... *Boracic* ...... *Boron*
.... *Nitric* ...... *Nitrogen*
.... *Muriatic* ...... *Chlorine*
.... *Succinic* ...... *unknown*
.... *Mellitic* ...... *unknown*

The bases of the four first, being metals, are
included in the 27 already adverted to; those of
the two next are the bases also of combustible substances, and therefore some description of the first
six bases will be given in their proper places; but
it will be requisite to give some account of phosphorus, fluorine, boron, nitrogen, chlorine, and of
the succinic, and mellitic acids; as well as of
water, hydrogen, and oxygen.

The necessity for including all these in the cata-
logue of the constituents of mineral substances will
become apparent as we proceed.

In the following list, therefore, are comprehended, according to the present state of our knowledge, the whole number of the

ELEMENTS OR ACCESSARY CONSTITUENTS OF
MINERALS.

Oxygen,
Hydrogen,
Water
Phosphorus
Fluorine
Nitrogen
Chlorine
Boron
The succinic acid
The mellitic acid
9 Earths
3 Alkalies
27 Metals
Carbon
Sulphur

Many of the substances included in the foregoing list are esteemed to be simple elementary bodies, because they have not hitherto yielded to any of the numerous attempts of the chemist to decompose them; others have only been partially analyzed, though sufficiently to determine that they are compounds; of others again the composition is known: others have altogether eluded the vigilant eye of the analyst.

a 5

Chemistry, notwithstanding the rapid advances that have been made in it, during the last few years, is still acknowledged to be far from perfect as a science. New facts continually arise, which as continually tend to illustrate and to advance the science of mineralogy, which is yet in its infancy, and is dependent in a very important degree on the advancement of chemical science.

In the following pages is inserted a short sketch of the nature and properties of each of the substances included in the above list of the elements or accessary constituents of minerals ; which, it is presumed, will tend to throw some light on the actual state of our knowledge of mineralogy in so far as it is dependent on chemistry ; as well as upon the affinity and relative proportions which these substances bear towards each other as mineral constituents.

Some account of the acids generally is likewise given, as well as of the earths, alkalies, metals, and combustibles : observations on each earth, alkali and metal, are inserted preceding the descriptions of such substances as are placed under each of them, in conformity with our present object.

OXYGEN.

Oxygen has not been obtained in a complete state of separation : in the most simple form in which it has been procured, it is combined with caloric, forming what is termed oxygen gas ; thus united, it is essential to the support of animal life.

Oxygen gas may be obtained from many sub-

stances; it is most abundantly, and perhaps most
readily, procured from the black oxide of manganese; which furnishes all the oxygen used by the
chemist, and all the oxygen used in the preparation
of the oxymuriatic acid consumed in the bleacheries
of Britain and other countries.

All the substances from which it can be procured,
are considerably diminished in weight after yielding oxygen gas, which is rather heavier than common air: all bodies which absorb oxygen acquire
an addition to their weight.

Oxygen was formerly considered to be the general
cause of acidity; in other words, a necessary principle of every acid; and the term Oxygen is compounded of two Greek words, having allusion to
that supposed theory: but that theory has lately
been done away, by direct proof of its not being
correct in two instances, which is further corrobo-
rated by the probability of its incorrectness in some
others; and that certain bodies afford acids by com-
bining with hydrogen.

Oxygen, it is ascertained, is so abundant a principle in many minerals, particularly of those constituting the oldest and most plentiful masses of the
crust of the globe, that it may be said to be one of
the most common and most abundant of mineral
elements, if not the most common and most abundant of all.

Of the most plentiful of all mineral substances,
silex, it forms 54 per cent.; of alumine 46; of
lime 28; of magnesia 38; of potash 17, and of
soda 26 per cent.: to which it may be added, that

a 6

it forms about 88¼ per cent. of water; and that in
the ores of tin and manganese, and many of those
of iron, lead, copper, &c. oxygen enters as an ingredient in various proportions.

Oxygen also forms an important ingredient in
many minerals, as an essential element in certain
acids; as in the two abundant substances the sulphate and carbonate of lime. It has been supposed
that the latter alone constitutes one-eighth part of
the whole crust of the globe. It may be assumed
that limestone is composed of 56 parts of lime and
44 of carbonic acid. Now lime consists of about 72
per cent. of calcium, and 28 of oxygen; and carbonic acid of about 28 per cent. of carbon, and 72
per cent. of oxygen; so that oxygen enters into
the composition of the one-eighth part of the crust
of the globe, which is calculated to be constituted
of carbonate of lime, in point of fact, nearly in the
proportion of one-half.

But argillaceous rocks are considered to be more
universal and plentiful than calcareous, and siliceous more abundant still. Of these rocks oxygen
forms on an average 50 per cent.: so that the cal-
culation in regard to the proportion in which oxygen enters into the composition of minerals, would
amount to a very large percentage of the whole
crust of the globe.

HYDROGEN.

The most simple form in which Hydrogen has been obtained, is that of a gas, in which it is in union with caloric, or the matter of heat. It is considered to be an elementary body.

Hydrogen is one of the component elements of water; its name is compounded of two Greek words, importing that circumstance; it is one of the elements of sulphur, and also, as it is believed, of phosphorus, of ammonia, and of the fluoric, and muriatic acids. It is obtained, in variable proportion, from several of those substances, which are termed combustibles; and, in combination with sulphur, forming sulphuretted hydrogen, it has been detected by analysis, in the Haüyne or Latialite; the swinestone or stinkstone, a variety of carbonate of lime which is found in considerable abundance, is supposed to owe the peculiarly offensive odour which it gives out when scraped or rubbed, to the presence of sulphuretted hydrogen.

Hydrogen gas is emitted from the crevices of volcanic matter; and it is asserted by Brongniart, that near St. Barthelemi, which is not far from Grenoble in France, hydrogen gas issues from the crevices of a country which has no appearance of being volcanic; and consisting of a grey friable argillaceous schistus. The gas has no odour; and if inflamed, continues burning sometimes for many months : the surrounding mountains are calcareous. He likewise says, that similar circumstances occur in England, on the road between Warrington and Chester, and also near Brozely in Shropshire.

WATER.

Water is composed of oxygen and hydrogen, in the proportion of about 88¼ of the former to 11¾ of the latter.

Water may be considered as merely an accessary, and not as an element, in some minerals : it is occasionally enclosed in crystal and chalcedony, and in variable proportion in certain minerals of a granular or loose texture ; but, in some others, it is an essential principle, as is evinced by the difference existing between the forms of the primitive crystals of the common, and of the anhydrous, sulphate of lime : the latter of these is composed of lime and sulphuric acid ; the former, of lime, sulphuric acid, and 21 per cent. of water : when water is an essential principle, it is termed *water of crystallization*.

Water is found in very different proportions, in a large number of earthy, as well as of metallic, minerals, both crystallized and massive.

The pure alkalies, potash and soda, retain even after fusion, about 1-5th of their weight of water ; and all acids, in a liquid state, contain water as an essential element.

ACIDS.

It is impossible to give such a description of the acids as will characterize them altogether. The greater part of them are chemically described as

found to be the *mineralisers* of several of the earths
and of the metals.

The metals and metalliferous ores are chiefly
found in veins, of which they occasionally com-
pose the only substance ; but they are more often
disseminated in veins, through earthy or stony
substances : such a substance is thence termed the
gangue or *matrix* of the mineral. Metalliferous
ores are less commonly found in masses or in beds :
a few of them occasionally occur imbedded in certain rocks.

They are met with in veins traversing almost
every kind of rock, but are most common in primitive and transition rocks, than in flœtz rocks :
they occur but sparingly in alluvial deposites, and
more rare in volcanic matter.

The comparative age of the metals is chiefly
judged of by the nature of the rocks which enclose
them. Iron and manganese have been detected by
every analysis in mica, a constituent of the oldest
primitive rock, granite ; tin and molybdena occasionally occur imbedded in it ; they also, as well
as tungsten, titanium, cerium, uranium, chrome,
and bismuth, are found almost exclusively in such
veins as traverse the oldest of the primitive rocks :
the foregoing metals may therefore be considered
of the earliest formation. Arsenic, cobalt, silver,
nickel, and copper, are presumed to be less ancient,
because though they occur in the oldest primitive
rocks, they are also found in newer. Gold, tellurium, and antimony, are considered to be metals
of a middle age, as they occur in the newer primi-

tive and the older secondary rocks. Lead, zinc,
and mercury, are found in the greatest quantity in
secondary formations, and are therefore supposed
to be less ancient than the preceding. Platina,
palladium, rhodium, iridium, and osmium, having
never been found *in situ*, it is impossible properly
to judge of their relative age ; but as crude platina
involves small portions of palladium, rhodium, iridium, and osmium, as well as of copper, gold, and
lead, we may conceive them to be of a middle age,
and shall there in the following series, place them
next to gold.

In respect of age, therefore, the metals may be
ranked as follows, and we shall accordingly, be-
gin the description of metalliferous ores with the
important ores of iron :

Iron
Manganese
Molybdena
Tin
Tungsten
Titanium
Cerium
Uranium
Tantalium
Chrome
Bismuth
Arsenic
Cobalt
Nickel
Silver

Copper
Gold
Platina
Rhodium
Osmium
Iridium
Palladium
Tellurium
Antimony
Lead
Zinc
Mercury

Iron is an ingredient in almost every rock, from the oldest primitive to the newest alluvial; and also in very many earthy and metalliferous minerals, and in all soils: it is therefore considered to be the most abundant and most generally diffused of all the metals. Wherever found, and with whatever combined, it is mostly in the state of an oxide, except when combined with sulphur.

Manganese, with iron, is an ingredient of mica, which is a constituent of the oldest granite; it occurs both in the primitive and secondary countries.

Molybdena may be reckoned a rare metal; it is occasionally found imbedded in granite, or in veins passing through it. It occurs only in the state of an acid or an oxide, or mineralized by sulphur.

Tin is abundantly and almost exclusively found in veins passing through primitive rocks, chiefly in granite and argillaceous schistus. Tin is always in

the state of an oxide : it occurs only in one compound metalliferous ore.

Tungsten is by no means a plentiful metal,
it usually accompanies tin : it occurs only as an
acid combined with iron, or as an oxide combined
with lime, in veins in primitive mountains.

Titanium occurs chiefly in the state of an oxide,
and may be reckoned a rare metal : it is usually
combined with iron, sometimes with silex.

Cerium is an extremely rare metal.

Uranium is also rare : it occurs chiefly in the
state of an oxide in primitive veins.

Tantalium is still more rare : it occurs in the
state of an oxide ; in one of its ores it is combined with iron, in the other with the rare earth
Yttria.

Chrome is a scarce metal, and occurs only in
the state of an acid, mineralizing lead and iron.

Bismuth is not a common metal ; it occurs in
the native state, also mineralized by sulphur, and
combined in some of the ores of silver, and of cobalt.

The preceding metals, being chiefly found in the
oldest primitive rocks, are considered to be of the
earliest formation ; the succeeding five are supposed
to be less ancient, because they occur both in the
oldest primitive and in certain of the secondary
rocks.

Arsenic is a more abundant metal than most of
the preceding : it is involved in small portions in
several of the native metals, in all the ores of cobalt,
and in most of those of silver.

Cobalt is not found alloying any metal ; in its ores it is combined with iron and arsenic ; it is not plentiful.

Nickel is a rare metal : it occurs as an oxide, and also combined with arsenic.

Silver is a somewhat abundant metal ; and occurs in greater or less quantity in most mineral countries : in the native state, it occurs in veins and beds, and disseminated in rocks : its ores are numerous ; it occurs combined with lead, copper, iron, antimony, tellurium, gold, quicksilver, and arsenic, and mineralized sulphur, and by certain acids.

Copper is an abundant metal ; it occurs in the native state : its ores are numerous, and in them copper is combined with iron, sulphur, silex, oxygen, and certain acids : it occurs in most mineral countries.

The three following metals are found in the newer primitive and older secondary rocks, and therefore are metals of a middle age.

Gold, though less abundant than silver, is more so than most of the preceding, and is not to be esteemed a rare metal ; though occasionally met with in veins, it is chiefly found in rivers and alluvial deposites : it occurs from 1 to 26 per cent. in the ores of tellurium, and sometimes in small portions alloying the native metals, copper, antimony, platina, and arsenic.

Tellurium is a rare metal : it occurs in the native state, but mostly is alloyed by a little gold : in its ores it is combined with gold, silver, lead, copper, and sulphur : it has only been found in two or three places.

Platina is not a plentiful metal : it is found only in certain districts in America, and only in the native state ; alloyed by small portions of gold, lead, copper, iron, osmium, iridium, and rhodium.

Palladium is rare ; it is found with platina, in the native state, alloyed by small portions of platina and iridium.

Iridium and Osmium are also found accompanying platina, together forming an alloy ; they also alloy platina, and the former of them, palladium : they are both rare.

Rhodium is found only alloying the platina of Peru, and is therefore extremely rare.

Antimony is not a very rare metal : it occurs in the native state, alloyed by small portions of iron and silver : in its ores it is combined with sulphur, silex, and oxygen : it occurs in few mineral districts.

Lead may be considered as the most abundant and most universally diffused metal after iron : it never is found in the native state, but its ores are very numerous : it occurs abundantly mineralized by sulphur, and by certain acids ; and is found in the state of an oxide : it occurs in certain ores of tellurium.

Zinc is not a scarce metal, but is pretty generally diffused : in its ores, it occurs combined with sulphur, iron, and silex.

Mercury is found only in a few places, but is not scarce : it occurs native, and combined with silver, sulphur, and with certain acids.

COMBUSTIBLES.

Combustibles form, in the mineral kingdom, a class of substances, having peculiar properties, and by no means agreeing amongst themselves in internal or external characters, and differing essentially from the earths, the alkalies, and the metals. Combustibles include both the hardest and the softest of mineral substances.

Several of the combustibles are found in a liquid state, some of these are translucent and even transparent ; but the greater number are solid ; when solid they are easily broken ; they possess neither the opacity, brilliancy, nor the weight of metals, being rarely more than twice the weight of water ; some of them are lighter than water.

If we were to class among combustibles, all those bodies whose chemical characteristic is that they will endure combustion, we should err, because many of the metals have that character.

Most of the metals whose properties are altered by combustion, acquire an increase of weight thereby ; whereas combustible substances are sensibly diminished in weight by the same process. The product of some of them is liquid, of others, solid; if solid, it is insoluble in water. Combustibles are either simple or compound.

The mineral bases of combustible substances may be said to be only two, viz. carbon and sulphur.

The purest form in which *Carbon* is seen, is that of the diamond ; and it was for a long time considered that the only chemical difference between this gem and charcoal is, that the latter contains some oxygen, and therefore is an oxide of carbon. But the late experiments of several chemists, and particularly of Sir H. Davy, tend to shew that there is no oxygen in pure charcoal ; and that there is no decided chemical difference between it and the diamond. Charcoal, however, always contains either hydrogen or water in very small and variable proportions, but not as an essential ingredient : the diamond is absolutely free from hydrogen and water ; and it is in this respect only, and in the mechanical arrangement of its particles, that there is any evidence of its differing from charcoal. The experiments of Allen and Pepys tend to prove that the actual quantity of carbon, in equal weights of diamond and charcoal, is precisely the same.

Carbon forms the basis of several of the combustibles, as coal, bitumen, amber, &c. ; and it enters into the composition of a few minerals in small proportion ; in the Aberthaw limestone, the hepatite, semi-opal and in clay slate, not exceeding 1 or 2 per cent. ; in rotten stone 10 per cent. ; and less than 1 per cent. in compact manganese : its

most important mineral character is, that it forms
the base of the carbonic acid, which enters into all
limestone rocks, as an ingredient, in the proportion
of about 44 per cent : carbonic acid consists of
about 28 per cent. of carbon and 72 of oxygen.

Sulphur was suspected by Berthollet to contain
hydrogen, and this suspicion has since been confirmed by Sir H. Davy during the career of his
brilliant discoveries, by means of the application of
the astonishing powers of galvanism or electricity to
many bodies which heretofore were considered to
be simple or elementary. Some experiments of
the same able chemist, tended to evince the presence of oxygen as another ingredient of sulphur :
but the later experiments by Guy Lussac have
proved that oxygen does not enter into its composition.

Sulphur is not only itself a highly inflammable
body, but is also an ingredient of other combustibles; as of certain kinds of coal. Large deposites of
sulphur are met with in some volcanic countries :
it is found in considerable masses or in beds, both
in primitive and transition countries ; and it is
largely involved in certain minerals ; such as iron,
copper, lead, antimony, silver, &c. which thence
are termed sulphurets of those metals, and which,
generally speaking, are the most abundant of all
metalliferous ores ; and it is met with in one earthy
mineral, the Fahlunite, in the proportion of 17 per
cent. Sulphur is the base of the sulphuric acid ;

which consists of 40 per cent. of sulphur and 60 of
oxygen. The sulphuric acid enters largely into the
composition of that abundant substance, sulphate
of lime or gypsum ; and is likewise an ingredient of
several other earthy minerals ; and in certain metallic ores.

EXPLANATIONS OF TERMS

Commonly used in Mineralogical Description.

———

Acicular. Long, slender, and straight prisms, or crystals, are termed acicular, from the latin, acicula, a little needle.

Aggregated. A mineral or rock is said to be aggregated, when the several component parts only adhere together, and may be separated by mechanical means: the felspar, quartz, and mica, constituting granite, may be separated mechanically. Granite is an aggregated rock.

Alliaceous. The odour given out by arsenical minerals, when exposed to the blowpipe or struck by the hammer, resembles that of garlic; in latin, allium; whence alliaceous.

Amorphous. Without form; of undefinable shape; from the Greek, αμορφος (amorphos) having that signification. Amorphous minerals are sometimes described as being of indeterminate, or indefinite forms.

Anhydrous, from the Greek ανυδρος (anudros), signifying without water: anhydrous gypsum is without water.

Arborescent. From the Latin, arboresco, to grow like a tree: see Dendritic.

Botryoidal. From the Greek, βοτρυωδης (botruodes) signifying, hung with clusters of grapes or berries. So a mineral presenting an aggregation of large sections of numerous small globes, is termed botryoidal; but when the globes are larger, and the portions are less, and separate, the appearance is expressed by the term mamillated. These forms may be observed in certain ores of cobalt, copper, and manganese, and often in chalcedony.

Bladed. This term relates chiefly to the structure of such minerals as, on being broken, present long flat portions, somewhat resembling the blade of a knife ; this appearance may in general be considered as the effect of interrupted crystallization.

Brittle. This character of mineral bodies does not depend upon their hardness ; those of which the particles cohere in the highest degree, and are immovable one among another, are the most brittle. The diamond, quartz, sulphate of barytes and sulphur, vary greatly as to hardness ; they are all brittle.

Canaliculated ; presenting deep channels on the surface, resulting either from interrupted crystallization, or the aggregation of numerous crystals.

Capillary, is derived from the Latin, capillus, a hair, and is chiefly used to express the long, tortuous, hair-like appearances, to be observed in native gold, and silver, and some other minerals Crystals are sometimes termed capillary, when long and slender ; but being mostly straight, they are more properly designated by the term acicular.

Cavernous. A mineral in which there are considerable hollows or cavities, is said to be cavernous.

Cellular. This term is used by Werner in the description of such minerals as exhibit cells formed by the crossing and intersecting of the laminæ or lamellæ of which they are constituted : commonly, any mineral presenting numerous small cells or cavities, is termed cellular : see vesicular.

Chatoyant, has been adopted from the French, who use it to express the changeable light resembling that to be observed in the eye of a cat, to be seen in certain minerals ; as in the Cat's eye.

Cleavage. This term is most commonly used in relation to the fracture of those minerals which, having natural joints, possess a regular structure, and may be cleaved into geometrical fragments ; as, into varieties of the parallelopiped, the rhomboid, &c.

Coherent. This term relates to structure. In minerals that are brittle, the particles are strongly coherent; in such as are friable, they are slightly coherent.

Columnar distinct concretions, is commonly used to express the great and small columns in which certain basalts and iron ores are found : but Werner includes under this term all the columnar appearances in every mineral consisting of numerous aggregated crystals, which readily divide into long and narrow portions of irregular form, owing to interrupted crystallization—such as the amethyst, pyrites, fluor spar, quartz, &c.

Compact, is a term which relates wholly to structure : and is that in which no particular or distinct parts are discernible ; a compact mineral cannot be cleaved or divided into regular or parallel portions. The term compact is too often confounded with the term massive.

Conchoidal, relates only to fracture ; and is doubtless derived from the Latin, conchoïdes, signifying like the shell of a fish. Fragments of many of the brittle minerals exhibit this appearance, and occasionally in great perfection, as quartz and sulphur; the fracture of compact minerals is frequently more or less perfectly conchoidal.

Concretion, generally signifies a small and distinct mass.

Coralloidal, resembling branches of coral.

Cuneiform, wedge-shaped ; cuneus, in Latin, signifies a wedge. This term relates only to fracture.

Decomposed. This term, when used strictly in a mineralogical sense, imports the decomposition which takes place naturally in any substance. Certain ores of iron, &c. in which sulphur pre-

dominates in an unusual degree, decompose by
exposure to air.

Decrepitate. A mineral is said to decrepitate on
exposure to heat, when it flies with a crackling
noise similar to that made by salt when thrown
into the fire.

Dendritic; derived from the Greek, δενδριτις (den-
dritis) signifying, like the growth of a tree. The
terms arborescent and dendritic are used synonimously : they are alike applied to the tree-like
appearance in which native silver and native
copper are sometimes found; to the delineations
seen on the surfaces of certain minerals; and to
the appearance in the mocha-stone, &c.

Dentiform, or *Dentated;* in the shape of teeth ; dens
being the Latin for a tooth.

Disseminated. When a mineral, whether crystallized or otherwise, is found here and there imbedded in a mass of another substance, it is said
to be disseminated in the mass.   Crystals of
quartz sometimes occur, disseminated in Carrara
marble, &c.

Disintegrated. This term is generally used to express
the falling to pieces of any mineral, without any
perceptible chemical action.

Diverging, or *Divergent*, relates to the structure of a
mineral.   When the structure is fibrous, and the
fibres are not parallel, they usually diverge in
part, but not wholly, around a common center;
as in certain zeolites, and hæmatites iron ores.

Drusy, has been adopted from the German term dru-
sen, for which we have no English word.   The
surface of a mineral is said to be drusy when
composed of very small prominent crystals, nearly
equal to each other; it is often seen in iron pyrites.

Elastic. A mineral which, after being bent, springs
back to its original form, is elastic.   Mica is
elastic ; talc, which greatly resembles mica, is
only flexible.

Earthy. This term relates to structure. Chalk and certain of the ores of iron and lead are notable instances of it.

Fasciculated. When a number of minute fibres or acicular crystals occur in small aggregations or bundles, they are said to be fasciculated ; a term doubtless derived from the Latin, fascis, a bundle. This appearance often occurs in green carbonate and arseniate of copper.

Fibrous. This term relates both to form and structure. Certain minerals, as amianthus, amianthiform arseniate of copper, a variety of gypsum, &c. occur in distinct fibres. Asbestus, gypsum, red hæmatites iron ore, &c. are found massive, and of a parallel fibrous structure : some varieties of red hæmatites and other minerals are of a radiating fibrous structure ; and the fibres diverge from a common center.

Filament. A mineral is said to occur in filaments, when it is found in slender, thread-like or hair-like portions. It is therefore nearly synonymous with the term capillary.

Filiform, is used in the same sense as the preceding ; but Werner confines its use to express the appearance of certain metals which occur in the form of wire, as native silver and native copper. Filum in Latin, signifies thread ; filum metalli, wire.

Fistuliform. Minerals occurring in round hollow columns, are termed fistuliform ; fistula, in the Latin, signifies a pipe. Stalactites and iron pyrites occur fistuliform.

Flexible. Talc is flexible ; it readily bends, but does not return of itself to its original form. Mica is both flexible and elastic.

Foliated. This term, which doubtless is derived from the Latin foliatus, having, or consisting of leaves,

is used by Werner to express the structure of all
minerals that may be divided or cleaved regularly, and are therefore by him said to consist of
folia or leaves. The structure of such minerals is
more commonly expressed by the term lamellar;
and they are said to consist of lamellæ or laminæ.

Fracture, is a term now chiefly employed in desig-
nating the appearance of minerals which have no
regular structure, when they are broken : such
minerals present an even, uneven, or a conchoidal fracture, &c.

Frangible. The term frangibility has relation to the
susceptibility of minerals to separate into fragments by force : this quality in minerals is not
dependent on their hardness ; the structure of
some and the brittleness of others, renders them
easily frangible ; while others, which from their
softness, and the ease with which their particles
or molecules yield or slide over one another, are
much more difficulty frangible ; such minerals
possess the character of toughness. Quartz is
easily broken, Asbestus is tough.

Friable. A mineral whose portions or particles slightly
cohere, and which is therefore easily crumbled
or broken down, is said to be friable, or in a friable state.

Fungiform. Certain substances, as for instance calcareous stalactites, are occasionally met with having a termination similar to the head of a fungus;
whence they are said to be fungiform.

Gangue, Gangart. We have these terms from the
Germans ; the gangue of a mineral, is the substance, in, or upon which, a mineral is found:
it is sometimes termed the matrix. Silver, occuring in, or upon carbonate of lime, is said to have
carbonate of lime for its gangue or matrix.

Geode. This also we derive from the Germans. A
geode is a hollow ball; at Oberstein in Saxony

are found hollow balls of agate lined with crystals of quartz or amethyst, which are termed geodes.

Glance is also a German word meaning shining; thus we have glance-coal, copper-glance, &c.

Globular distinct concretion is used to express the character of any mineral which occurs in little round or roundish masses; the Pea-stone and Roe-stone are examples of it.

Granular. The structure of a mineral is said to be granular, when it appears to consist of small grains or concretions; which sometimes can, sometimes cannot, be discerned without the help of a glass; we have therefore the fine granular, and the coarse granular structure.

Greasy is used in relation to lustre: fat quartz has a greasy lustre.

Hackly. This term relates to a fracture which is peculiar to the malleable metals; which, when fractured, present sharp protruding points.

Hæmatites is derived from the Greek αιματιτις, signifying blood-red; it was first applied by mineralogists to the variety of iron ore which is now called the Red Hæmatites; but has since been extended to other iron ores of the same structure, but differing in colour. We have also brown hæmatites, and black hæmatites iron ore.

Hepatic. A term derived from the Latin, hepar, the liver; it is applied either to colour or form. We have hepatic pyrites, hepatic quicksilver; the hepatite.

Hydrate is derived from the Greek ὕδωρ, (udor) water; and is applied to certain of those minerals (as the hydrate of magnesia) of which water forms an ingredient in very large proportion.

Imbedded. A mineral found in a mass of another substance, is said to be imbedded in it. Crystallized quartz occurs imbedded in Carrara marble.

It also occurs partly imbedded in other substances, as in fluor.

Indeterminate. Indefinite. These terms are used synonimously with Amorphous in describing minerals which have no particular or definable form. Crystals of which the form cannot be accurately ascertained, are said to be of indeterminate forms.

Incrusting : any substance covered by a mineral, is sometimes said to be incrusted by it : thus the various articles which are placed for a certain length of time in certain springs or wells in Derbyshire, &c. and which are by some supposed to be converted into petrifactions, are only incrusted with calcareous, or argillaceous matter.

Interlacing. Interlaced. When the fibres or crystals of a mineral are found intermingling with each other in various directions, they are interlacing, or interlaced.

Investing. A mineral spread upon, or covering another, is sometimes described as investing it.

Irridescent. This term relates only to the various colours with which the surfaces of some metalliferous minerals are naturally tarnished : as yellow copper ore, iron pyrites, galena, sulphuret of antimony, &c.

Lamellar ; this term relates to structure : when a mineral can be fractured or cleaved into regular and parallel plates or laminæ, its structure is said to be lamellar ; and the portions thus obtained are termed laminæ or lamellæ ; these terms have been adopted from the Latin, in which they were almost synonimously used to express thin plates of any substance.

Lamellar distinct concretions. This term is sometimes used to express the structure of certain minerals (as the oxide of uranium) consisting of laminæ which cohere but slightly.

Lamelliform. A mineral consisting of lamellæ, is said to be lamelliform.

Laminæ, *Lamellæ*.　See *Lamellar*.

Lenticular is employed to express the forms of certain crystals which are nearly flat, and convex above and beneath ; and which consequently resemble a common lens.

Malleability.　Some of the metals suffer extension when beaten with a hammer ; and are therefore termed malleable metals. Native gold and native silver are very malleable metals.

Mamillated.　See *Botryoidal*.

Massive.　This term is sometimes used in describing a substance of indeterminate form, whatever may be its internal structure ; but is more commonly used in contra-distinction to the term crystallized, as applied to those minerals which possess regular internal structure, without any particular external form.

Matrix.　See *Gangue*.

Meagre.　This term relates to the touch or feel of a mineral. It belongs chiefly to some of those minerals which are of an earthy texture. Chalk is remarkably meagre to the touch.

Natural joints.　Such minerals as can be broken into regular forms, as the cube, rhomboid, &c. can be cleaved into those forms, only in the direction of, or along, their natural joints. In some minerals however, which have not yet been regularly cleaved, the natural joints are perceptible by the assistance of a strong light.

Nacreous relates to lustre ; and is employed to express the lustre of some minerals (as of the pearl spar) which greatly resembles that of pearl. Nacre de Perle, in French, signifies Mother of Pearl.

Nodular.　A mineral which presents irregularly globular elevations, is termed Nodular. Flint is found in nodular masses.

Opake. Those minerals are opake which do not transmit a perceptible ray of light even through the thinnest and smallest pieces.

Pass into. One mineral is said to pass into another, when both are found so blended in the same specimen, that it is impossible to decide where the one terminates, and the other begins. Flint is found passing into chalcedony.

Pectinated. If a mineral exhibit short filaments, crystals, or branches which are nearly equidistant, it is pectinated: pecten, in Latin, signifies a comb.

Porous. A mineral is said to be porous, when it is traversed in different directions with communicating holes which pass through the substance.

Pseudomorphous. Minerals exhibiting impressions of the forms peculiar to the crystals of other substances are said to be pseudomorphous. Quartz exhibiting crystals in the form of the cube; calamine, such as are peculiar to carbonate of lime, &c. are termed pseudomorphous: ψευδος, in Greek, signifies false; μορφη, form or figure: sometimes they are termed secondary crystals.

Pulverulent. When the particles of a mineral are very minute and cohere very slightly, or not at all, it is said to be pulverulent; or in the pulverulent state.

Radiated; radiatus, in Latin, signifies beset with rays; when the crystals of a mineral are so disposed as to diverge from a center, they are said to be radiated.

Ramose; ramus, in Latin, signifies the branch of a tree; a mineral having that appearance is described as being ramose.

Refractoriness. This term is used both chemically and mechanically in relation to minerals. It is sometimes applied to those which strongly resist the application of heat; and occasionally to some whose toughness enables them to resist repeated blows.

Reniform. Kidney-shaped; ren, in Latin, signifies kidney.

Retiform, Reticulated. Minerals occuring in parallel fibres, crossed at right angles by other fibres which also are parallel, exhibit squares, like the meshes of a net. Retis, in Latin, signifies a net. We have reticulated native silver, native copper, red oxide of copper, &c. And it may be remarked that such minerals as occur reticulated, generally assume the cube, as one of their crystalline forms.

Scopiform. If a number of minute crystals or fibres are closely aggregated into a little bundle, with the appearance of diverging from a common center, they are said to be scopiform. Scopa in Latin, signifies a broom or besom.

Schistose structure. Minerals which split only in one direction, and present fragments which are parallel, but of unequal thickness, which also are not smooth and even, and are without lustre, are said to possess a schistose structure. Schist in the German signifies slate.

Sectile. This term relates to structure, and is derived from the Latin, seco, to cut. Those minerals are termed sectile which are midway between the brittle and the malleable. A slice or portion cut from a sectile mineral, is fragile, and the new surface on the mass is smooth and shining. Plumbago and the soapstone are both sectile.

Semi-transparent. A mineral is said to be semitransparent when an object is not distinctly seen through it.

Slaty-structure. This term is synonimous with Schistose-structure, which see.

Specific Gravity. The specific gravity of minerals is determined by comparison. The usual mode of

determining it is by weighing them in pure distilled water; the weight of which is assumed to be
1 or unit. Earthy minerals vary from twice, or less
than twice, to nearly five times the weight of distilled water. Metalliferous ores and native metals
vary from five to seventeen times its weight: some
minerals, especially some of the combustibles,
are lighter than water, and are of course supernatant.

Specular Minerals are those which present one smooth
and brilliant surface which reflects light. We
have specular red iron, specular iron pyrites, &c.
These are said always to occur close to the walls
or sides of veins. Speculum, in Latin, signifies
a looking glass.

Spicular and Splintery Fracture belong to minerals
of an imperfectly crystalline form. These frac-
tures do not greatly differ: they are both irregular; the spicular is shorter and more pointed than
the splintery.

Stalactitiform. σταλάγμα, (stalagma) in the Greek,
signifies a drop, an icicle. Stalactitiform minerals
greatly resemble icicles in shape.

Stellated. When the structure of a mineral is fibrous,
and the fibres diverge all round a common center,
its structure is said to be stellated: stella, in
Latin, signifies a star.

Striæ, Striated. The slight channels occasionally observable on the planes of crystallized minerals
are termed striæ, and the crystals on which they
are seen are said to be striated. The striæ are
commonly parallel and generally indicate the
direction in which crystals may be cleaved.
Stria, in Latin, signifies a groove, or channel.

Structure. This term relates to the internal characters of minerals. Such as can be cleaved into
regular forms, presenting smooth, brilliant, and
parallel surfaces, are said to have a crystalline
structure; but when the surfaces are neither

smooth nor parallel, and when, on the contrary they are rough and curved or undulating, the structure is said to be imperfectly crystalline; under which term also may be comprehended all fibrous minerals whether massive or not. All such as have no determinate structure, as those minerals which are granular, splintery, &c. or compact, may be included under the term indefinite or promiscuous structure.

Supernatant. Such minerals as are lighter than water, and consequently swim upon it, are said to be supernatant. Supernato, in Latin, signifies to swim or float upon.

Tabular. When this term is used in relation to structure it is nearly allied to the schistose or slaty. Talc, mica and roofing slate are described by Werner as possessing a tabular structure. This term sometimes is used to express the external form of crystals : such as are nearly flat, and whose length and breadth are nearly the same, are sometimes called tabular crystals; from the Latin, tabula, a table or board.

Toughness relates to internal structure. Those minerals which are bruised, or suffer depression, by repeated blows in the attempt to fracture them, are esteemed to be tough.

Translucent. A mineral through which an object cannot be seen, but which transmits some light, is termed translucent. Rock salt, sometimes quartz, flint, and fluor, &c. are translucent : many minerals are translucent on the edges, as common marble, &c.

Transparent. Those minerals are transparent through which an object may be clearly seen.

Tubercular. A mineral whose unevenness of surface arises from small and somewhat round elevations, is said to be tubercular. Flint is sometimes tubercular.

c 3

Tuberous: exhibiting somewhat circular knobs, or elevations.

Tubular: see fistuliform.

Vesicular. A mineral is said to be vesicular, when it has small and somewhat round cavities, both internally and externally. Lava, pumice, limestone, basalt, &c. are sometimes vesicular: from the Latin, vesicula, a little bladder.

Vitreous; from the Latin, vitreus, glassy; minerals having the lustre of glass, are said to possess the vitreous lustre.

Unctuous. The term relates to the touch. Pipe clay is somewhat unctuous: Fullers' earth is unctuous; plumbago and soapstone are very unctuous.

CONTENTS.

This Table shews the order in which the Minerals comprehended in the following pages have been described. The whole number of ingredients in each compound mineral are not noticed in this table. In SILEX, &c., the &c. relates to the small portions of oxide of iron, oxide of manganese, or water, which many of them contain. The complete analysis is included in the description. ———

EARTHY MINERALS.

CONTENTS. EARTHY MINERALS.

CONTENTS. EARTHY MINERALS.

CONTENTS. EARTHY MINERALS.

CONTENTS. NATIVE METALS, &c.

CONTENTS. NATIVE METALS, &c.

CONTENTS. NATIVE METALS, &c.

d

CONTENTS. NATIVE METALS, &c.

In the heads, or running Titles, of some of the ensuing pages, the following contractions have been requisite.

a. ac.	for	*acid*
Alum.	—	*Alumine*
bit.	—	*bitumen*
c.	—	*chrome*
car. carb.	—	*carbon*
carb. ac.	—	*carbonic acid*
i. ir.	—	*iron*
Mag.	—	*Magnesia*
m. man. mang.	—	*manganese*
o. ox.	—	*oxygen*
Pot.	—	*Potash*
So.	—	*Soda*
sul. hyd.	—	*sulphuretted hydrogen*
sul. acid	—	*sulphuric acid*
sulph.	—	*sulphur*
w. wa. wat.	—	*water*

EARTHY MINERALS.

Including such as principally consist of one or more of the Earths, either nearly pure, or combined with an alkali, or an acid, or with small proportions of certain metallic oxides, or with water, &c.

SILEX.

This Earth is, when pure, in common with the rest of the earths, perfectly white and infusible, except by the intense heat of voltaic electricity. It has neither taste nor smell, and its specific gravity is 2.66.

Silex has never been found mineralized by any acid, but is occasionally involved in small proportion in some of the acidiferous earthy substances; it forms a large ingredient of very many earthy minerals, including some of the hardest gems and the softest clays; it is proved by analysis to enter, in variable proportion, into the composition of about two-thirds of the whole number of earthy minerals whose composition is known; and as it is the chief ingredient of the oldest and most plentiful of the primitive rocks, and is found in rocks of almost every age and formation, it is esteemed to be the most abundant substance in nature.

Silex, as well as the rest of the earths, has lately been proved, by Sir H. Davy, to be a compound substance; it consists, according to Berzelius, of oxygen, in the proportion of about 54 per cent, united with a base, *Silicium*, which has not hitherto been obtained in a state of separation, in the proportion of about 46 per cent. Silex cannot therefore be now considered as a simple or elementary body.

Notwithstanding the complete analysis of silex, it still obtains among chemists its old denomination of an Earth; principally, it may be supposed, from the difficulty of properly characterizing its base; which is not believed by Sir H. Davy to be a metal, but of a peculiar nature, bearing an analogy to boron, charcoal, sulphur and phosphorus.

As common flints are almost wholly composed of *siliceous earth,* it thence received the name of Silex, which in the Latin signifies flint; but it is found in the greatest purity in quartz or rock crystal.

QUARTZ.

Quartz is found crystallized, fibrous, granular, and compact. It scratches glass, does not yield to the knife, and is infusible. Its specific gravity is 2.6; and it is composed of silex, with 2 or 3 per cent. of water.

Crystallized quartz is found perfectly transparent and colourless; also, red, yellow, grey, black, brown, purple, green, and of various shades of each colour.

The transparent and colourless is known by the name of Rock Crystal: the largest and most esteemed crystals are brought from Madagascar, the Alps, Norway, and Scotland; where they are found in cavities in granite. Single crystals have been

met with, of more than 100 lbs. weight. These are
bought at a high price by the lapidary, to cut up
into various ornaments, as seals, &c. and into proper forms for spectacles, as a substitute for glass.
In smaller crystals, quartz is found in almost every
metallic vein, both of ancient and recent formation,
in every kind of rock.

Quartz, more often than any other crystallized
mineral, contains foreign substances ; sometimes
drops of water, with bubbles of air, may be seen in
it ; also crystals of schorl or titanium, crystals of
chlorite, and iron ore.

The crystallizations of quartz or rock crystal are
very interesting. The crystals in my possession
exhibit 40 distinct varieties of form ; the most common of which is a hexahedral prism terminated by
hexahedral pyramids : the two pyramids joined base
to base, without an intervening prism, are rarely
seen. The primitive crystal is also rare, but is
occasionally found in the neighbourhood of Bristol ; it is an obtuse rhomboid, very nearly approaching the cube. Its angles, according to Haüy,
are 94° 24' and 85° 36' ; but the results obtained
by the reflecting goniometer do not correspond
therewith.

Quartz allows, though not readily, of mechanical
cleavage, parallel with the planes of its primitive
crystal.

The transparent crystals found in the neighbourhood of Bristol, termed *Bristol Diamonds*, are
crystallized quartz ; those of Cornwall are by some
called *Cornish Diamonds*.

Crystals of quartz of a light yellow, or of various shades of brown, are brought from many places. The best are found in a hill called Cairn-gorm, in Scotland. A single crystal about twelve inches long, and four in diameter, of a deep brown colour and transparent, which was fit for the lapidary, was not long since sold by public auction for 210 guineas. These are by some called *False Topazes.*

When of a reddish purple, or violet colour, quartz is called *Amethyst* : the crystals are generally of the deepest colour towards the summit. It commonly occurs in veins in metalliferous mountains in Spain, Bohemia, Saxony, Hungary, &c. never in those of primitive granite. Frequently it is found in hollow masses, called geodes, which are occasionally surrounded by a coating of agate; but these are principally met with in volcanic countries; in Auvergne, the Tyrol, and the Palatinate. Analysis has proved the amethyst to contain a very minute quantity of iron and manganese, to which its colour may be attributed. Amethyst has been met with in the tin mines of Cornwall, Polgooth and Pednandrae.

Fibrous quartz is yellowish or greyish white, and occasionally pale amethyst: sometimes it occurs in radiated and globular concretions, two inches or more in diameter, but only in Cornwall.

Granular quartz is white, yellowish or greyish white; it occurs in granular distinct concretions, sometimes in mass, and as a component of certain granites. It is fine, or large grained. The fine grained, with silvery mica, composes a granite near Schihallien, in Scotland; the larger grained forms large blocks in argillaceous schistus and other rocks in Scotland. The latter becomes snow white by calcination, and is largely employed in the porcelain manufactory.

Compact quartz is of various colours, and occurs in mass, or disseminated, or globular, &c. It is found entering into the composition of rocks, from the oldest to those of the most recent formation, and composing veins and beds in others : sometimes it is found in considerable blocks, though it seldom forms entire mountains.

Prase is of a leek green colour, and translucent ; it occurs in mass at Brutenbrun, in Saxony, in a mineral bed : it appears to be an intimate mixture of quartz and actinolite.

When of a light grass or an apple green, and somewhat transparent, quartz is termed *Chrysoprase,* which is found in mass, imbedded in serpentine, in Siberia, with opal, chalcedony, &c.

Avanturine is yellowish red, or grey, greenish, or blackish. It appears to be filled with silvery and yellowish spangles, that reflect light with great brilliancy. Some suppose these spangles to be mica, others imagine that the appearance is produced only by the particular direction of the laminæ. It takes a good polish, and is used for seals and other ornaments. The best avanturine is brought from Spain.

A variety of quartz, which is commonly massive, and has a greasy lustre, as though it had been rubbed with oil, is therefore called *Fat quartz :* it is one of the gangues of native gold in Peru.

Another variety is opaque white, and is thence

termed *Milk quartz*. It has sometimes a tinge of
red, which often passes into a beautiful rose red;
when it is termed *Rose quartz;* it often has the
greasy lustre of fat quartz. Its colour is said to be
owing to manganese. It has been found at Raben-
stein in Bavaria, in considerable quantity, in a vein
of manganese traversing a large grained granite. It
has also been met with in Finland, and near Cork
in Ireland.

Quartz sometimes exhibits impressions of the crystalline forms of substances on which it has been deposited, but which have been decomposed; quartz
exhibiting such crystalline appearances is termed
Pseudomorphous. Sometimes it is merely cellular;
and when the cavities are very minute, and the
quartz is in very thin plates which intersect each
other in every direction, it is so light as to swim on
water; whence this variety has been termed *Swim-
ming stone*, by some *Spongiform quartz*. It has
been found at Schemnitz in Hungary, at Joachimstal in Bohemia, at Schneeberg and Freyberg in
Saxony, at Beresof in Siberia, and in Cornwall in
England. The cavities of one specimen in my possession, from Pednandrae Mine, near Redruth, are
partly filled up with fluor spar, the external parts of
which are rounded, shewing it to be in a state of
decomposition; the cavities of another, from Relis-
tean Mine, are filled up partly by black, partly by
bright yellow copper ore.

Quartz combined with variable proportions of iron,
is termed *Ferruginous quartz:* it is of a yellow or
red colour, and opake, and is found both compact

and crystallized. It is harder than pure quartz; and when heated, becomes magnetic. It is sometimes met with in remarkably neat small crystals having both terminations perfect, and of a yellowish or reddish colour. These crystals have been principally found in secondary rocks, near Compostella in Spain; whence they are called *Hyacinths of Compostella.* Massive ferruginous quartz, or *Eisenkiesel,* is found in the veins of primitive mountains, where it is often met with as the gangue of various metallic substances, as of lead, copper, sulphuret of iron, and sometimes of gold.

A variety termed *Hyalite,* or *Muller's glass,* having in many respects the appearance of chalcedony, has been found in small masses upon, or lining the cavities of, amygdaloid. It bears a striking resemblance to gum-arabic, and is said to be composed of 92 parts silex and 7 of water. It has been found only in volcanic countries: in Tuscany,—in small stalactites in the rocks of Piperino; in Solfatara, &c.

Quartz is sometimes found forming beds, and more often veins, in primitive mountains. The quartz in these veins is sometimes compact, but is occasionally hollow in places; in these cavities the crystals which are seen in the cabinets of mineralogists are found. It occasionally occurs imbedded, as in porphyries; and in remarkably neat transparent crystals, in Carrara marble.

It is also met with in veins or caverns in secondary countries of different natures; and forms a large proportion of alluvial deposites, principally in fragments,

or rounded or angular grains, constituting sand;
which is sometimes, by causes which we know not
how to explain, found adhering, forming masses denominated *Sandstone* and *Gritstone.*

OPAL.

Opal is either of a clear, or of a bluish white: it
includes several varieties.

It is found in small masses or in veins in Hungary,
in rocks which seem to be in part decomposed, and
which are by some considered to be volcanic; by
others, as argillaceous rocks, the result of the decomposition of porphyries. In these rocks both the
common and the noble opal occur.   Opal is also
met with in Iceland, and Saxony.

The *Common opal* is usually white with a tinge of
yellow, red, or light green, internally. It consists of
93.5 parts of silex, 1 of oxide of iron, and 5 of
water.   It has been found in several of the mines
of Cornwall.

The *Noble opal* exhibits changeable reflections of
the same colours as the former variety, and is an exceedingly brilliant and beautiful mineral : it is hard
enough to scratch glass.   The finest specimens of it
are in the Imperial Cabinet of Vienna; one is about
5 inches long and $2\frac{1}{2}$ in diameter, the other is of the
shape and size of a hen's egg.   It consists of 90
parts of silex, and 10 of water.

Semi-opal is harder than the preceding varieties
and is mostly opake ; occasionally transparent, with
a glistening resinous lustre.   It is principally met

with in secondary countries; sometimes in volcanic rocks, and in basalts. It has also been found in primitive granite and porphyry, especially in the veins traversing those rocks which contain silver. It consists of 85 per cent. of silex, 1 of carbon, 1.75 of oxide of iron, 8 of ammoniacal water, and a small portion of bitumen. Semi-opal is found in Auvergne in France, in the island of Elba, in Bohemia, Iceland, Hungary, &c. Fossil teeth have been found penetrated by this mineral.

Wood-opal has a ligneous structure, and is met with of various shades of grey, brown and black. It appears to be wood, penetrated by opal or semiopal; and is found near Schemnitz, and at Telkobanya in Hungary.

A variety, met with at the same place as the preceding, called *Ferruginous opal* or *Opal jasper*, is of a yellowish or yellowish brown colour, with a glistening resinous lustre. It consists of about 43 parts of silex, 47 of oxide of iron, and 7 of water.

HYDROPHANE.

The Hydrophane is considered to be a variety of opal. It is generally whitish, and nearly opake; by immersion in water, it exhibits some of the changeable colours of the former varieties, and is found in the same places. It consists of about 93 parts of silex, 2 of alumine, and 5 of water.

Hydrophane is porous, and commonly adheres to the tongue. It is chiefly found in Saxony, the Isle of Ferroe, and in Hungary. At Mussinet near Turin, it occurs in veins of chalcedony, or of hard serpentine, traversing a serpentine mountain in every direction.

The Menilite is by some considered a variety of semi-opal. Its common colour is a smoke brown; its structure slaty; it is somewhat translucent, and is found in irregular masses in beds of clay, between beds of sulphate of lime at Menil-montant near Paris. It is sometimes called the *Pitchstone of Menil-montant.* It consists of 85.5 parts of silex, 1 of alumine, 11 of water and inflammable matter, with small portions of lime and oxide of iron.

Flint is of various shades of white, yellow, brown and black, and is somewhat harder than common quartz; it is readily broken in any direction, and has a conchoidal fracture and a glimmering lustre. It is found in irregular masses, and sometimes forming the substance of certain marine organic remains, as echinites and coralloids; and consists of 97 parts of silex, 1 of alumine and oxide of iron, and 2 of water. Its specific gravity is 2.58.

Flint is said occasionally, though rarely, to be found in veins in primitive rocks; but it is also said that the flint thus found has not precisely the characters of common flint. It is met with in nodules in compact carbonate of lime in Derbyshire; at Mont-martre near Paris, in an impure sulphate of

lime; it is also found in certain marls; but that
which may be termed its ordinary native place, is
the upper chalk formation, in which it is met with
in regular layers, and occasionally, as in Fresh-
water Bay in the Isle of Wight, in continuous beds
of considerable length. Flint is also abundantly
found in portions evidently rounded by attrition,
forming deposites in the neighbourhood of chalk hills,
and of gravel in alluvial countries. When red, yellow or brown, they are termed *Ferruginous flints.*

The formation of the flints which lie in detached
masses, though in parallel layers, in chalk, has
much occupied the attention of geologists, and
without producing any satisfactory solution. Their
form proves that they have not been rolled, or con-
veyed into the chalk; in which they bear every appearance of having been formed: they are frequently found containing shells. Some naturalists have
ventured on the supposition that the places in which
they are found were formerly occupied by animals;
and that the formation of flints in those places, has
been owing either to the affinity existing for silex in
the animal matter, or that it has been converted
into flint. This it must be obvious is mere hypo-
thesis, and is not more deserving of regard than the
notion entertained by others, that, contrary to all
the known principles of chemistry, these flints have
been formed by the conversion of lime into silex by
some unknown natural agency. The most commonly
received opinion seems to be, that flints have been
formed by the filtration of siliceous matter through
the chalk; a theory not without serious difficulties.

Kirwan quotes from Schneider's Topog. Mineral.
114. that 126 silver coins were found enclosed in
flints, at Grinoc in Denmark; and an iron nail at
Potsdam.

CHALCEDONY.

Chalcedony is found of various shades of white,
yellow, brown, green, and blue. It occurs massive;
forming veins; in round balls, termed geodes; and
also, botryoidal and stalactitical; sometimes it bears
the impression of organized bodies; it is frequently
met with coating crystals of quartz, and occasionally
in cubic crystals, which, it is ascertained, are only
secondary, or pseudomorphous. It is commonly
semi-transparent; it has no regular fracture, and is
harder than flint. Its elementary constituents have
not been accurately ascertained, but as it is often
found passing into flint, it may reasonably be assumed, that their analysis would not greatly differ.
The specific gravity of chalcedony is about 2.6.

Chalcedony is found principally at Oberstein in
Saxony, and in the isle of Ferroe. A blue variety
is met with in Transylvania. That of Iceland is
in thin layers, alternately more or less translucid,
and perfectly parallel. But the most superb specimens were brought from a copper mine in Cornwall,
called Trevascus, which was situated in argillaceous
schistus: these specimens are translucid, whitish,
and variously ramified. A variety of a beautiful
blue colour on the surface was found in a tin vein
passing through granite, in Pednandrae mine, near

Redruth, in the same county; and another variety, very much resembling flint, containing small portions of yellow copper, in Relistian mine. Chalcedony has also been met with in several others of the Cornish mines. It was anciently procured from Chalcedon in Upper Asia, whence its name.

When of a white colour and translucent, chalcedony is called *Cacholong.* Some varieties are opake and adhere to the tonge. Cacholong is found with chalcedony, sometimes with flint. It is met with at Champigny, near Paris, in a calcareous breccia: but the real cacholong is found in the banks of the Cach, a river in the country of the Calmucs of Bucharia.

When chalcedony contains appearances of arborization, or vegetable filaments, which have been supposed to be owing to the infiltration of iron or manganese through its natural crevices, it is termed *Mocha-stone.* This is believed chiefly to be brought from Mocha in Arabia.

Carnelian is of various shades of yellow, brown, and red. It is found in several places in Europe; but the most valuable specimens are brought from Arabia, and from Surat and Cambay in India, where it is said to be found in certain rocks in the globular, or the stalactitical form.

Sard is supposed to differ from the preceding variety only in its colour, which is orange-yellow, passing into brownish yellow.

When yellowish, white, red, or yellow, brown, or brownish black, and opake, it is called *Jasper.*

When two or three of the above varieties are mixed in alternate and concentric bands, exhibiting, when cut and polished, zones, or angular lines like fortifications, the compound is termed *Agate*. It is found in the form of irregular rounded nodules from the size of a pin's head to more than a foot in diameter, or in veins or strata, or occasionally stalactitic. Sometimes agate is found in amygdaloid and in gypsum ; near the Wolga it occurs between strata of secondary limestone.   The most beautiful agates of Britain are found in the neighbourhood of Perth and Dunbar, and are called *Scotch pebbles* ; but the most celebrated are those of Oberstein, in Saxony. When the colours are disposed in straight parallel bands, it is called *Ribbon agate*, by some *Ribbon jasper*. Another variety represents rocks or buildings, and is called *Ruin agate*.

But if two or more of the preceding varieties are associated with a band of milk-white opake chalcedony, the mass is called *Onyx*. Onyxes cut into portions about the size of a bean, exhibiting opake white circles, resembling the iris of the human eye, are termed *Onyx-eyes*.

Sard, united with opake milk-white chalcedony, is called *Sardonyx*.

Plasma seems to be a variety of chalcedony, which it resembles in being translucent, and somewhat harder than quartz.   It is of a dull greenish colour, with yellow and whitish dots, and has a glistening lustre.   It has not been analyzed.   Plasma is brought from Italy and the Levant ; and is said to occur at Taltsa, in High Hungary : also, dis-

seminated in rounded pieces, with ffint and hornstone, in a mountain of serpentine, at Bojanowitz,
in Moravia. Its specific gravity is 2.04.

Heliotrope is mostly of a deep green colour, and
translucent ; and commonly, yellow or blood-red
spots are interspersed through the substance. From
the latter circumstance it has obtained the name of
Bloodstone. It is considered to consist of chalcedony coloured by chlorite, or by green earth ; and
is found in Siberia, Iceland, and in a vein at Jasch-
kenberg, in Bohemia,—but the most beautiful varieties are brought from the east ; whence, among
lapidaries, by whom it is in considerable request, it
has obtained the name of *Oriental jasper.* The specific gravity of heliotrope is 2.6.

CIMOLITE.

Cimolite is of a light greyish-white, inclining to
pearl-grey, but by exposure to air it acquires a reddish tint; it occurs massive, and of a somewhat
slaty structure ; is opake ; yields to the nail, and
adheres to the tongue. It often encloses small
grains of quartz. It consists of 63 parts of silex,
23 of alumine, 1.25 of oxide of iron, and 12 of
water. Its specific gravity is 2.

It abounds in the island of Cimola (whence its
name) now called Argenteria, situated near that of
Milo. It was employed by the ancients, and still
is by the inhabitants of the island, for some of the
purposes to which fuller's earth is applied.

BLACK CHALK. DRAWING SLATE.

This mineral is of a greyish or bluish black co-lour; it has a slaty structure, is meagre to the touch, and soils the fingers. It is found in primitive moun-tains, accompanying argillaceous schistus, particu-larly that which is aluminous, to which it is nearly allied; and is said to occur occasionally in the neighbourhood of coal formations. It is met with near Pwllhelli, in Caernarvonshire; in Isla, one of the Hebrides; in France, Spain, Italy, Iceland, and in Bareith. That from the latter place yields by analysis about 64 parts of silex, 11 of alumine, 11 of carbon, with small proportions of iron and of water.

ALAMANDINE.

This mineral, commonly termed *Noble,* or *Pre-cious garnet,* is usually of a brilliant crimson colour, sometimes with a slight smoky tinge, and transparent. It occurs granular, and also crystal-lized, in some of the forms usual to the common garnet; its structure is imperfectly lamellar, though but rarely visible. Its specific gravity is 4.3; and it consists of 35,75 parts of silex, 27,25 of alu-mine, 36 of oxide of iron, and 0.25 of oxide of manganese.

The alamandine is very much esteemed as a precious stone. The most beautiful, which are sometimes of reddish violet colour, are brought from Sirian, the Capital of Pegu : among lapidaries, they are improperly called Syrian garnets. They appear to be the *Carbuncle* of the ancients. Of their geological situation in Pegu, we are entirely ignorant. They are also found in Bohemia, Hungary, Ethiopia, Madagascar, Brazil, &c.

In Bohemia, they are found near Meronitz and Trziblitz, in the circle of Leutmeritz, disseminated in an alluvial deposite, consisting principally of fragments of serpentine and rounded masses of basalt, cemented by a grey marl. In this deposite are also found hyacinths, beryls, sapphires, quartz, magnetic iron, and even fossil shells.

It is said to have been met with in granite at Strontian; also at Ely in Fifeshire, and at Wicklow, in Ireland.

TABULAR SPAR.

This very rare mineral has only been found at Oravitza, and at Dognaska, in the Bannat of Temeswar, entering into the composition of a rock, consisting principally of bluish carbonate of lime and brown garnets; or, according to others, in a vein of bluish lamellar carbonate of lime, containing green garnets.

The tabular spar is of a greyish white colour ; translucent and somewhat hard. It is phosporescent when scratched with a knife, and is said to have been met with in six-sided tables, in which the natural joints may be discovered, parallel to the sides of a slightly rhomboidal prism. It is composed of 50 parts of silex, 45 of lime, and 5 of water. Its specific gravity is 2.86.

JENITE. YENITE.

The Yenite is a scarce mineral which at first sight greatly resembles hornblende. It is of a brown, or brownish black colour, dull externally, and of about the hardness of felspar. It occurs amorphous, acicular, and crystallized; generally in the form of a rhomboidal prism : six varieties have been discovered in the form of its crystals, the primitive of which is a rectangular octohedron, measuring over the summit, according to Haüy, one way, $112^{\circ}. 36'$, the other, $66^{\circ}. 58'$. It is composed of 29 parts of silex, 12 of lime, and 57 of the oxides of manganese and of iron ; and therefore does not seem to belong to earthly minerals ; but it is always ranked in that class. Its specific gravity is 3.8.

It has been found only in Corsica, at Rio la Marine and Cape Calamite ; it is dispersed in crystals and almost compact round masses, in a thick bed of a green substance, the nature of which has not been determined, but which bears a considerable analogy to the Yenite ; and is accompanied by yellowish-green epidote, quartz, and arsenical iron. The bed containing the yenite lies upon another, consisting of large grained carbonate of lime, enclosing talc. At Cape Calmite it is accompanied by magnetic iron ore, granite, and quartz.

STEATITE.

Steatite is of various shades of white, grey, yellow, green, and red; and is met with massive, and (at Bareuth) with occasional appearances of internal crystallization; which, being mostly, if not always, referable to the forms assumed by quartz or carbonate of lime, are therefore in varieties which cannot originate in the same primitive form, and are thence considered to be only pseudomorphous.

This substance has generally a very unctuous and soapy feel, but it differs from the soapstone in not having alumine as one of its elementary ingredients: it yields to the nail, but does not adhere to the tongue; its fracture is splintery, sometimes slaty. It considerably resembles some varieties of serpentine, but is much softer. The grotesque figures brought from China, are generally supposed to be a variety of steatite; which, though it possesess some characters in common with the substance of which those figures are made (the Agalmatolite), differs essentially in respect of analysis. Steatite consists of 64 silex, 22 magnesia, 3 oxide of iron, and 5 of water. Its specific gravity is 2.67.

Steatite is found in considerable masses, or in beds or veins, in some primitive mountains. It is most common in serpentine. At Freyberg in Saxony, it is met with in tin veins; where it is accompanied by, or mingled with, mica, asbestus, quartz, and occasionally native silver, &c. It is found at Portsoy, in Scotland, in serpentine, and in the Isle of

Sky, in wacke. It occurs also in the Isle of Angle-sey. It abounds in the principality of Bareuth, in Saxony, Bohemia, Norway, Sweden, and France.

The white varieties, or those that become so by calcination, are employed in the manufactory of the finest porcelain : other varieties are said to be used for fulling. The Arabs, according to Shaw, use steatite in their baths instead of soap, to soften the skin ; and it is confidently asserted that the inha-bitants of New Caledonia either eat it alone, or mingle it with their food. Humboldt says, that the Otomaques, a savage race inhabiting the banks of the Oronoko, are almost entirely supported during three months in the year, by eating a species of steatite, which they first slightly bake, and then moisten with water.

BRONZITE.

The colour of this mineral is brown, having fre-quently the aspect of bronze ; its structure is fibrous lamellar, and its lustre is considerably metallic. It consists of 60 per cent. of silex, 27. 5 of magnesia, 10. 5 of oxide of iron, and 0. 5 of water. Its specific gravity is 3, 2.

The Bronzite is usually found disseminated in Serpentine. It occurs in the Col de Cervière in the Alps ; at Matray in the Tyrol ; at Basta in the Duchy of Wolfenbutel ; at Dobschau in Upper Hungary, &c. : It seems to belong to primitive countries.

LAUMONITE.

This mineral occurs in aggregated crystalline masses, deeply striated, or in separate crystals of several varieties of form; the primitive of which, according to Haüy, is a rectangular octohedron, or, according to Bournon, a rhomboidal tetrahedral prism, with rhombic bases. The Laumonite is white, transparent or translucent, and hard enough to scratch glass. It was formerly termed the *efflorescent zeolite*, on account of its undergoing a spontaneous change by exposure to the air; in consequence of which it loses its natural transparency, and becomes opake, tender, of a shining white colour, and pearly lustre; eventually, it falls into a white powder similar to that resulting from the decomposition of Glauber's Salts. It has lately received the name of the Laumonite, from Gillet De Laumont, who first made known its true nature. It specific gravity is 2, 2. It is composed of 49 of silex, 22 of alumine, 9 of lime, 17. 5 of water, and 2. 5 of carbonic acid.

This mineral was first discovered in the lead mine of Huelgoet in Brittany, lining the cavities of the veins. It has since been brought from Ferroe; from near Paisley in Renfewshire; from Portrush in Ireland, and Larne in the Isle of Sky.

DIPYRE.

This rare substance occurs in slender octohedral
prisms, of a greyish, or reddish white colour, fasciculated into masses. These prisms exhibit joints
parallel to the sides, and to the diagonal, of a rectangular prism. The Dipyre is of a shining vitreous
lustre, is hard enough to scratch glass, and becomes slightly phosphorescent by the application of
heat. It consists of 60 of silex, 24 of alumine,
10 of lime, and 2 of water. Its sp. gr. is about
2, 7.

It was found in the torrent of Mauleon, in the
western Pyrennees, in a gangue of white, or reddish steatite, mingled with sulphuret of iron.

STILBITE.

Stilbite is of a peculiar glistening or shiny pearly
lustre, by which it may be recognized at once. Its
colours are white, grey, brown, or red; it is
transparent or translucent, and not sufficiently hard
to scratch glass. It occurs lamelliform, massive,
and crystallized; the crystals are sometimes fasciculated into slender prisms: the form of the primitive
crystal is a right prism with rectangular bases, in
which it sometimes occurs, but more often these
prisms are terminated by tetrahedal summits. It
consists of 52 parts of silex, 17. 5 of alumine, 9
of lime, and 18. 5 of water. Its sp. gr. is 2. 5.

The Stilbite is met with in the fissures of primitive rocks; in mineral veins; and in the cavities of amygdaloid.

It has been found near Grenoble, of a pale straw colour; at Andreasberg, upon carbonated lime; at Arendahl, in Norway, of a brown colour; in Iceland, of a shining white colour, on the Iceland spar; in the islands of Skye, Staffa and Canna, in amygdaloid; and it has been met with massive, of an orange brown colour, at Dumbarton; at Glen Farg in Perthshire, and at Callhill in Aberdeenshire.

CAT'S EYE.

This mineral is generally brought in the polished state from the coast of Malabar, and from Ceylon; but nothing is known of its geological situation. Its colour is of various shades of grey, green, brown, or red; and it exhibits a peculiar play of light, resembling the eye of the animal from which it takes its name; this peculiar reflection, emphatically termed by the French, chatoyant, is owing to the fibrous texture of the substance, arising, as it has been supposed by some, from its consisting of asbestus, enclosed in quartz. Its specific gravity is about 2.7. It is composed of 95 parts of silex, 1.75 of alumine, 1.5 of lime, and 0.25 of oxide of iron. It is frequently employed as a precious stone, and is in considerable estimation.

PREHNITE.

This mineral is of a pale greenish or yellowish colour, with a shining pearly lustre, and is some-what transparent; it is scarcely hard enough to scratch glass, and becomes electric by heat. It occurs in very minute crystals, which are for the most part closely aggregated; their primitive form is a right rhomboidal prism of 103°. and 77°; the prisms sometimes have 6 or 8 sides. A variety which occurs in small translucent lamellæ, of a yellowish white colour, and glistening pearly lustre, consists of 48 parts of silex, 24 of alumine, 23 of lime, and 4 of oxide of iron. The phrenite is also found massive, and consists of the same elements, somewhat differing in their respective proportions, together with about 2 per cent. of water.

Crystallized phrenite has been met with in con-siderable quantity, and of a purer green than that of Europe, at the Cape of Good Hope; it occurs in France; in the valley of Fascha in the Tyrol, ac-companying mesotype; and at Dunglasse in Scot-land.

The lamelliform variety, called the *Koupho-lite,* occurs near Barèges, and the peak of Eredlitz in the Pyrennees, in a gangue of cavernous horn-stone, mingled with chlorite, &c.

The massive is found in France; in Scotland, near Dumbarton; at Hartfield Moor, near Paisley; at Frisky Hall, near Glasgow; at the Castle Rock, near Edinburgh; and in the Isle of Mull.

The Prehnite seems to belong to rocks of early formation, of which it is does not enter into the composition; it is only disseminated in small quantity, without forming either beds or veins.

ZOYSITE.

The Zoysite, which, together with the Thallite, is included by Haüy under the name of EPIDOTE, was so called after the Baron de Zoys. It occurs in oblique rhomboidal prisms, of a grey, greyish yellow or brown colour, with a pearly lustre and translucent, which are rarely perfect, owing to deep longitudinal striæ. The Zoisite consists of 44 parts of silex, 32 of alumine, 20 of lime, and 2.5 of oxide of iron, and is met with in Carinthia, the neighbourhood of Salzburg, and in the Tyrol, &c.

IDOCRASE. VESUVIAN. BROWN VOLCANIC HYACINTH.

Idocrase occurs massive, but more often crystallized in groups, consisting of short quadrangular prisms, of which the edges are variously replaced, and variously terminated. Haüy notices eight varieties in the form of its crystals; one of which, if complete, would have presented 90 planes; 16 on the prism, and 37 on each summit; he considers the primitive to be a right prism, with square bases, differing very little from the proportions of the

B

cube. The colour of the Idocrase is mostly brownish or yellowish green, sometimes orange, with a
resinous lustre ; and it is hard enough to scratch
glass.   That of Vesuvius consists of silex 35.50,
alumine 33, lime 22.25, and oxide of iron 7.50.
Its specific gravity varies from 3.088 to 3.409.   It
possesses double refraction.

It has been met with both in volcanic, and in
primitive countries.   It occurs in the midst of the
projected masses of Vesuvius and Etna ; where its
crystals, which exhibit no appearance of change by
heat, line the cavities of volcanic rocks, principally composed of felspar, mica, talc, and carbonated lime ; and are accompanied by garnet, hornblende, meionite, &c.

The Idocrase has also been found in Siberia, in
a greenish white serpentine, near the lake Achta-
ragda, and on the banks of Vilhoui ; and in massive veins passing through green serpentine, in the
plain of Mussa in Piedmont.   It has been found
also in the counties of Wicklow and Donegal, in
Ireland.

It is cut and polished by the lapidaries of Naples, under the name *Crysolite of Vesuvius.*

GARNET.

The Garnet is of a reddish, yellowish, greenish,
or blackish brown colour ; it is found in small granular masses, and crystallized in the form of the
dodecahedron with rhomboidal planes, which is

considered to be the form of its primitive crystal. It also occurs in crystals having 24 trapezoidal faces; only 5 or 6 varieties have been described. It is harder than quartz, but not so hard as the Almandine. It is rarely transparent, frequently opake. It consists of 43 parts of silex, 16 of alumine, 20 of lime, and 16 of oxide of iron.

The almandine, allochroite, melanite, aplome and garnet, are commonly arranged together under the latter name; but their elementary constituents do not correspond.

Garnets are very abundant; they principally occur disseminated among some of the older rocks, as micaceous schistus, serpentine, and gneiss, and sometimes in granite. They are met with in most countries in which those rocks occur, and sometimes are so plentiful as almost to constitute the mass. They are found also in mineral veins, accompanying some of the ores of copper, lead, magnetic iron, mispickel, &c. In the mountains separating Stiria and Carinthia, they are met with upwards of 2 lbs. in weight, in a bed of green talc. In Bohemia they are found of a brown colour, accompanying tin; in Siberia of a pale green, in lithomarga; at Topschau, in Hungary, of an emerald green, in serpentine: in Corsica, of a yellow colour; in the Grisons, &c. of an orange colour; in Cornwall, in small quantity, in argillaceous schistus: they are not uncommon in Scotland, in micaceous schistus; and are found in some parts of Ireland.

CINNAMON-STONE.

This rare mineral has only been met with at Columbo, in the island of Ceylon. It is known in Holland by the name of Kanelstein, signifying cinnamon-stone, probably from its resemblance to cinnamon in colour. Its geological situation is not known. It occurs massive, or in detached fragments, which are full of cracks, and usually of a yellowish brown, passing into orange yellow and hyacinth red. It is somewhat transparent, with a vitreous-resinous lustre, and scratches quartz, though with some difficulty. By analysis it affords 38.8 of silex, 21.2 of alumine, 31.25 of lime, and 6.5 of oxide of iron. Its sp. gr. is 3.6. By some, it is considered as bearing considerable affinity to the Garnet.

TRIPOLI.

Tripoli obtained its appellation from being first brought from a place of that name in Africa; it has since been found in other places. This mineral has generally an argillaceous aspect. It is sometimes of a schistose structure, but is more often massive, with a coarse, dull, earthy fracture; and is meagre and rough to the touch, and yields easily to the nail. It occurs of various shades of grey, yellow, and red; and is said constantly to yield 90

parts of silex, the rest being argil, iron, and sometimes a small portion of lime.

Tripoli is found in beds at Menat near Resin, in
the Puy de Dome; in Tuscany, it it met with at
Volterra, so situated, as to seem the consequence
of the decomposition of chalcedony; and at Post-
Chappel in Saxony, in a mountain containing coal.
It is also found in Flanders, Westphalia, and
Russia.

It is used in polishing metals, marble, glass, and
other hard bodies.

BOLE.

Bole or *Ochre* is always somewhat compact; it is
either red, yellow, or brown : it yields to the nail,
adheres to the tongue, and gives a shining streak
on paper : when breathed on, it gives out a sensible
argillaceous odour; it breaks down in water, with
which it may be reduced to the consistence of a paste.

Red Chalk or *Reddle* is by some considered as a
variety of bole ; but from its containing a large proportion of iron, it has lately been placed among
its ores.

A variety, of a lighter red colour than red chalk,
is brought from Armenia, and is commonly known
by the name of *Bole armenic.*

Another variety, found in Lemnos, when impressed by the seal of the governor of the island, or
of the Grand Seignor, is sold under the name of
terra sigillata. It is used in medicine. *Red* bole
is found near Estremos in Portugal.

Bole of a *yellow* colour occurs in beds, between those of sand, and therefore belongs to the newest secondary formation. It becomes red by exposure to heat. It is met with at several places in France; that of Auxerre is composed of 65 of silex, 9 of alumine, 5 of lime, and 20 of oxide of iron.

The red brown earth of Sienna, used as a pigment, is considered to be a variety of bole.

Bole of a *brown* colour, or of a yellowish brown, is commonly known, as a pigment, by the name of *bistre*; it is found in the island of Cyprus, but nothing is known of its geological situation. It is used in porcelain painting.

Boles are considered as coloured marls or clays.

CLAY.

The substances comprehended under the term of Clay, may be generally described as any earthy mixture which possesses plasticity and ductility when kneaded up with water. Few, if any of the substances possessing these characters, can, strictly speaking, be considered as constituting a distinct mineral species; being, in the general, the result of the decomposition of rocks. Clay, when moist is plastic; somewhat unctuous to the touch, and acquires a polish by being rubbed with the finger or the nail; when dry, it is solid; when burnt, sufficiently hard to give sparks with the steel, and is infusible. Clays, generally speaking, have not been analyzed; though it is suspected that the proportions of their constituent principles vary considerably. They are considered in the aggregate to consist of a large proportion of silex, mixed with a third or fourth of their weight of alumine, occasionally with a small quantity of lime, a variable proportion of oxide of iron, and some water.

In the term Earthy Clay, may be comprehended common brick earth, or loam, and common alluvial clay.

Brick Earth or *Loam* varies very much in appearance, texture, and composition. It usually contains a considerable proportion of sand; which, nevertheless, is frequently added by the brick-maker. It is commonly met with above common alluvial clay, and frequently rests upon an interposed bed of sand. The organic remains contained in it are few.

Common alluvial Clay occurs principally in low countries, in which it serves to fill up hollows; it frequently rises into hills, which sometimes are stratified. In many countries considerable tracts consist principally of clay to a great depth, as in the London chalk basin, which has been perforated in some places to the depth of 500 feet, without passing through it; it contains the remains of large land and sea animals, and sometimes vegetable remains are found in the same bed; occasionally, at various depths, beds of fine white sand are met with. Deposites or beds of clay are considered to consist of the debris or ruin of rocks, and are regarded among the principal causes of the formation and duration of springs. The water which per-

colates the secondary country, bordering the deposites of clay, passes beneath them, and is retained
until some opening permits it to rise in the shape
of a spring, or until vent is given to these reservoirs
by the sinking of wells.

Pipe Clay is of a greyish or yellowish white colour, an earthy fracture, and smooth greasy feel;
it adheres pretty strongly to the tongue, and is very
plastic and tenacious. It is manufactured into tobacco pipes, and is the basis of the Queen's water
pottery.

Potters' Clay is plastic, slaty. It yields to the
nail; is generally of a reddish, bluish or greenish
colour, and has a soft, and often greasy, feel. When
mixed with sand, it is made into bricks and tiles.
A variety found in the forest of Dreux, in France,
employed, on account of its infusibility, in the making of tiles for the porcelain furnaces, consists of
43 parts of silex, 33 of alumine, 3 of lime, 1 of iron,
and 18 of water. Most part of the clay used in the
potteries of Staffordshire and Newcastle upon Tyne,
is said to be found near Teignmouth in Devonshire.
That of Hampshire yields by analysis, 51 parts of
silex, 25 of alumine, 3 of lime, with a trace of
manganese and some water.

Porcelain Clay is greyish or yellowish-white, or
more often reddish-white; it adheres to the tongue,
is meagre to the feel, and is commonly friable; but
if compact, is easily broken. It falls to pieces in
water, and becomes plastic, though not in a very
great degree. That which is found in considerable
beds in the parish of St. Stephen's in Cornwall,

consists, according to Wedgwood, of 60 parts of
alumine and 40 of silex. It seems undoubtedly to
originate from the decomposition of felspar; it frequently contains portions of quartz and of mica.
The origin of porcelain clay, in the general, is not
however well understood. It differs materially in
respect of composition. The *Kaolin* of China consists of 71.15 of silex, 15.86 of alumine, 1.92 of
lime, and 6.73 of water. It is found largely in
France, in granitic countries, and therefore seems
to have the same origin as that of Cornwall.

Indurated clay is met with interposed between
beds of coal at Stourbridge in Worcestershire, and
at Coalbrook Dale in Shropshire. It is sometimes
called *Stourbridge clay*, or *fire clay*. It occurs
massive, and in large compressed nodules of a greyish-white colour, with a tinge of blue or yellow;
it yields readily to the knife, and is very refractory
in the fire; by exposure to the air it becomes soft
and falls to pieces, and then becomes plastic.

PORCELLANITE.

Porcellanite is found massive, and of a slaty structure. Its colour varies from grey to bluish-grey,
mixed with red; ochry yellow; greyish or bluish-
black. It is opake, and hard enough to scratch
glass; but it is not abundant; being principally,
if not exclusively found in places in which mines of
coal have been in a state of combustion; and is regarded as shale altered by heat. At Mount Brus-

sant, near St. Etienne, in France, it is composed of layers, alternately grey and red ; that of Schlangenberg, in Bohemia, is of a dull green colour.

MELANITE.

The Melanite is usually black and opake, and occurs in the form of a rhomboidal dodecahedron, of which the edges are commonly replaced by planes. It consists of about 35 parts of silex, 6 of alumine, 32 of lime, 25 of oxide of iron, and a trace of oxide of manganese. The Melanite is usually arranged among garnets. It has been found in Italy, at Frascati, near Vesuvius, in a volcanic rock, enclosing also felspar, idocrase and hornblende: it also occurs in the calcareous rocks of Somma.

APLOME.

The Aplome is usually considered a variety of the garnet, with which it agrees in respect of its external figure, but differs in containing manganese. It commonly occurs in rhomboidal dodecahedrons, of which the planes are striated parallel with their lesser diagonal ; they are usually of a deep brown, or orange-brown colour, opake, and somewhat harder than quartz ; it consists of 40 of silex, 20 of alumine, 14,5 of lime, 14,5 of oxide of iron, and 2 of oxide of manganese. The aplome is found on the banks of the river Lena in Siberia. Garnets of a yellowish-green colour have been met with at Swart-zenberg, in Saxony, which have considerable affinity to this mineral.

THALLITE. ACANTICONITE.

This mineral is found granular, massive, or crys-tallized in six, eight or twelve-sided prisms, vari-ously terminated and longitudinally striated; and is of a green, yellowish, bluish or blackish-green colour, of a shining lustre, and somewhat transparent. The primitive crystal is a right prism, of which the bases are oblique-angled parallelograms. Haüy includes this mineral and the zoisite under the term *epidote*. The crystallized thallite consists of 37 parts of silex, 21 of alumine, 15 of lime, 24 of oxide of iron, and 1.5 of oxide of manganese; and its specific gravity in 3.45; but the granular variety, in the form of a green sand, varies in respect of the proportions of its elements.

The Thallite is not often found massive, but chiefly in crystals, varying in size from the acicular to near an inch in diameter; the acicular are met with in the department of Iseré, in France; at Chamouni, in the Alps, &c.; the larger occur at Arendahl, in Norway. It belongs exclusively to primitive rocks, but is only found in veins and fis-sures; magnetic iron, garnet, felspar, adularia, axi-nite, and asbestus, are the minerals which chiefly accompany thallite.

WERNERITE.

This mineral is of a greenish-grey or olive-green colour, with a lustre between pearly and resinous; it is softer than felspar, and yields to the knife. It occurs masssive, and in eight-sided prisms with four sided pyramids. It consists of 40 parts of silex, 34 of alumine, 16 of lime, 8 of oxide of iron, and 1.5 of oxide of manganese.

The wernerite is a rare mineral; it is met with in irregular grains or crystallized, disseminated in rocks composed of a greyish or of a red felspar, intermingled with quartz, at Bouoen, near Arendahl in Norway; in the mines of Nortbo and of Ulrica in Sweden, and at Campo-longo in Switzerland.

TOURMALINE.

Tourmaline in respect of colour, is either white, green, blue, brown, yellow or black; it occurs in crystals which are striated, or rather deeply channelled, lengthwise; their lustre is splendent or vitreous, and they are scarcely so hard as quartz. It is remarkable that this substance is either translucent or transparent when held up to the light, and viewed in a direction perpendicular to the axis of the prism; but if viewed perpendicular to the bases, is always opake, even though the prism be short. Seventeen varieties in the form of the crystal have been described; their primitive form is, according to

Haüy an obtuse rhomboid of 133°. 26′ and 46°. 34′.
The crystals become electric by being heated, and
thereby acquire polarity; and their summits or pyramids are always dissimilar. That which presents the
greatest number of faces always exhibits the positive
or vitreous electricity; and that having the smaller
number, always the negative, or resinous. Its spe‑
cific gravity is about 3.

The green tourmaline of Brazil is composed of
40 parts of silex, 39 of alumine, 3.84 of lime, 12.5
of oxide of iron, and 2 of oxide of manganese.

The white variety was found at St. Gothard in
micaceous dolomite by Dolomieu, who mentions
having discovered some in the granite of Elba, the
half of which was white, the other half black. The
yellow variety has been noticed in sand from Ceylon.
Tourmalines of a dull-green, or of a bluish-green,
are from Brazil; those of an emerald-green, from
Ceylon. A variety of an indigo-blue-colour, thence
called *Indicolite*, has been found in the mine of Utoe
in Sweden, in crystals of an indeterminate form,
disseminated in a gangue of steatite, quartz, and
felspar.

AXINITE. THUMERSTONE.

This mineral derived its name of Thumerstone,
from having been first met with at Thum in Saxony.
It occurs in lamelliform concretions, and crystallized. The lamelliform, of a dingy violet colour, is
found at Ehrenfriedersdorf in Saxony. The same
variety, of a dull clove-brown, is found at Botallack,

near the Land's End in Cornwall ; sometimes also
it is crystallized, though not very determinately.
The most beautiful is met with in a serpentine rock
at Balme d'Auris in Dauphiné ; generally in neat
and well defined crystals, sometimes nearly colourless and transparent, but more often of a dull reddish
violet colour and translucent ; whence it has obtained the name of *Violet Schorl of Dauphiné.* The
crystals in my possession exhibit 18 varieties of
form, which are not symmetrical. This want of
symmetry is common to those substances, which,
like the axinite, become electrical by exposure to
heat. The primitive crystal of the axinite is a remarkably flat right rhomboidal prism, of which the
bases are oblique angled parallelograms of $78°\frac{1}{2}$ and
$101°\frac{1}{2}$ according to Haüy ; but the measurements
obtained by the reflecting goniometer do not cor-
correspond with the results obtained by him. The
axinite is hard enough to scratch glass, but less hard
than quartz ; its specific gravity is about 3.2 ; and it
consists of 44 of silex, 18 of alumine, 19 of lime,
14 of oxide of iron, and 4 of oxide of manganese.

It has only been met with in the veins and fissures
of primitive rocks, and is not very abundant. Besides the places above mentioned, it occurs in the
peak of Eredlitz in the Pyrnnees, upon a gangue of
quartz, accompanied by carbonate of lime ; near
Alençon in granite ; at Mount Atlas, in Africa ;
near Kongsberg in Norway, in a white laminated
calcareous rock, accompanied by black mica, quartz,
and sometimes native silver.

ALLOCHROITE.

The Allochroite, is of a greyish, dingy yellow, or reddish colour, and opake; it is not so hard as quartz. It consists of 35 parts of silex, 8 of alumine, 30.5 of lime, 17 of oxide of iron, 3.5 of oxide of manganese, and 6 of carbonate of lime. It is commonly considered as a variety of the garnet, from which it differs in respect of composition.

The allochroite is found in the iron mine of Virums, near Drammen in Norway, accompanied by carbonate of lime, Hematites iron, and brown garnets.

LAPIS LAZULI.

This mineral is found massive, and of a fine azure blue colour; its texture is fine grained or compact, and it is hard enough to scratch glass, though it scarcely gives sparks by the steel. Its specific gravity is 2.76 to 2.94. Its blue colour is not uniform, as it frequently encloses iron pyrites, compact felspar, and quartz. It is said to have been met with crystallized in the form of a rhomboidal dodecahedron; but as the crystal was opake, and enclosed iron pyrites and carbonate of lime, there seems no sufficient proof of its being true Lapis Lazuli; which, according to Klaproth, consists of 46 of silex, 14 of alumine, 28 of carbonate of lime, 6.5 of sulphate of lime, 3 of oxide of iron, and 2 of water.

It has been found in small masses enclosed in

primitive rocks, principally in granites, accompanied by felspar, pyrites, garnet and carbonate of lime; but is more often found in small masses rounded by attrition; as on the borders of the lake Baikal in Siberia. The finest specimens are brought from China, Persia, and Great Bucharia.

Lapis lazuli is used in jewellery, but is chiefly important as affording that beautiful pigment called ultra-marine, so highly valued by painters on account of its great advantage of not changing by time and exposure.

EGYPTIAN JASPER.

This mineral is more commonly known by the name of *Egyptian Pebble.* It occurs in roundish masses which are externally rough, and generally of a brown colour. Internally it is usually of a light colour, which sometimes approaches to that of cream, around which are disposed irregular zones or bands of various shades of brown, sometimes intermixed with nearly black spots, and occasionally dendritic appearances. Its specific gravity is 2.5—2.6; by one analysis which does seem to have been complete, it yielded 75 parts of silex, 15 of alumine, and 5 of magnesia.

It is found, according to Dr. Clarke, in vast abundance, together with masses and detached fragments of petrified wood, among which are several varieties of the palm, scattered over the surface of the sandy desert, eastward of Grand Cairo, even to the borders of the Red Sea.

It is susceptible of a high polish, and is therefore often applied to ornamental purposes.

TREMOLITE.

The general colour of Tremolite is white, which sometimes has a greenish, bluish, yellowish or reddish tinge; it occurs fibrous, and crystallized in four, six, or eight sided prisms, terminated by diedral summits, and is semi-transparent or translucent, and hard enough to scratch glass. [Its specific gravity is about 3.; the fibrous variety of Clicker Tor in Cornwall is composed of 62.2 of silex, 14.1 of lime, 12.9 of magnesia, 5.9 of oxide of iron, and 1 of water.

It was first discovered in the valley of Tremola near St. Gothard, whence its name: it has since been met with in Hungary, Transylvania and Bohemia. In Corsica, it occurs in lamellar bluish green talc; near Nantes, in granite abounding in felspar; at Somma, in granular carbonate of lime; and in Bengal: in lamellar limestone, in the banks of the lake Baikal in Siberia, whence it has been called *Baikalite*; in Glen Tilt and Glen Egg in Aberdeenshire, in white primitive limestone; in Cornwall, it is found at Clicker Tor. A fibrous specimen in my possession from Stony Gwins in that county, is deposited on quartz, and accompanied by small yellow crystals of uranite.

MEERSCHAUM.

Meerschaum is of a whitish or yellowish white colour, opake and dull; it has an earthy fracture, yields easily to the nail, and adheres strongly to the tongue; sometimes it is so light as to swim on water, and occasionally is very porous; this last characteristic has doubtless occasioned its name, which signifies sea-foam. It consists of 50.5 per cent. of silex, 17.25 of magnesia, 0.5 of lime, 5 of carbonic acid, and 25 of water.

It occurs in the isles of Samos and Negropont in the Archipelago, in mass, or disseminated, or in beds: at Kiltschik in Natolia, it fills a vein about six feet wide, traversing compact grey carbonate of lime; it is soft when first dug, and in that state is made into pipes, but hardens by exposure to air. It is also met with in Carinthia.

In the Turkish dominions, Meerschaum is employed as fuller's earth is with us; and by the women as soap for washing their hair. In Constantinople it is termed Keffekil or earth of Kaffa, the town of the Crimea whence it shipped.

A substance somewhat similar to Meerschaum has been found at Cartel del Piano near Sienna, consisting of 55 of silex, 25 of magnesia, 12 of alumine, 3 of lime, and 0.1 of oxide of iron; it was made into bricks so light as to swim on water, thus restoring one of the lost arts recorded by Strabo and Pliny.

Another substance, consisting of 55 of silex, 22

of magnesia, and 23 of water, and of a chocolate
brown colour, is found at Salinelle near Sommières,
in beds, in chalk containing silex : and in various
places in Piedmont, a substance of a white colour,
consisting, when fresh dug, of silex, magnesia, and
water, is found in beds and in veins : by exposure to
air it absorbs carbonic acid.

ANTHOPHYLLITE.

The Anthophyllite has hitherto only been found
at Kongsberg in Norway; it occurs massive, with
joints parallel to the faces of a rectangular prism,
is feebly translucent on the edges, and has a slight
metallic lustre ; it is scarcely hard enough to scratch
glass. Its specific gravity is about 3.3. By analysis
it is found to consist of 62.66 of silex, 13.33 of
alumine, 4 of magnesia, 12 of oxide of iron, 3.25 of
oxide of manganese, and 1.43 of water.

Some of the characters of the Anthophyllite have
induced Haüy to suppose that it is only a variety of
the Hypersthene ; but their elements do not correspond.

HARMOTOME. CROSS STONE.

The harmotome is commonly met with in flattish
quadrangular prisms, terminated by four rhombic
planes, crossing each other lengthwise and at right
angles. It is also met with in solitary crystals.
Their primitive form, according to Haüy, is a rect-

angular octohedron of 86° 36′ and 93° 24′ ; but this
is not confirmed by the reflecting goniometer, which
gives results differing about 2°.   I possess crystals
of the harmotome in 12 varieties of form, one of
which is so remote from the primitive, as to appear
a perfect six-sided prism, and several approach that
form.   This transition is very intelligible, though
not easily described without figures.

In cruciform crystals it occurs in metalliferous
veins, mingled white lamellar carbonate of lime and
sulphuret of lead, at Andreasberg in the Hartz; it
is also met with at Strontian in Scotland.   In solitary crystals it is chiefly found in the cavities of
siliceous geodes at Oberstein in Saxony.

The colour of this mineral is greyish-white ; it is
translucent with a somewhat pearly lustre, and is
hard enough to scratch glass.   Its specific gravity
is 2.35 ; and it is composed of 49 per cent. of silex,
16 of alumine, 18 of barytes, and 15 of water.

ASBESTUS.

There are several varieties of asbestus.   They are
generally of a fibrous texture, varying in respect of
flexibility and elasticity.   The fibres of asbestus
have not yet been seen in any very determinate
form, but Haüy regarded some which fell under
his observation as rhomboidal prisms.   Asbestus is
extremely difficult of fusion in the mass ; but its
fibres are easily reduced by the blowpipe.   Asbestus
is derived from a Greek work, signifying imperish-
able.

Amianthus occurs in very long and extremely slender fibres, which are very flexible, and of a whitish, greenish or reddish colour. It consists of 59 per cent. of silex, 3 of alumine, 9 of lime, and 29 of magnesia.

It is found in the Tarentaise in Savoy, in the longest and most beautiful fibres : that of Corsica is less beautiful, but is so abundant, that Dolomieu used it for packing his minerals : near Barèges in the Pyrennees, it occurs mingled with felspar, lining veins passing through gneiss. It occurs also at Inverary and Portsoy in Scotland, and in the Isle of Unst.

Amianthus (signifying unsoiled) was woven by the ancients into a kind of cloth, in which, being incombutible, they wrapped up the bodies of their dead, before they were placed on the funeral pile, that their ashes might be collected free from ad-mixture.

Mountain Cork. The structure of this variety differs from the former; the filaments are not de-posited in a parallel direction, but intermingled in various directions, occasioning cavities, to which may be attributed the lightness of the mass. When in thin flexible plates, it is termed *mountain leather;* when in thicker and less flexible, *mountain cork.*

It occurs in the silver mines of Johan Georgen-stadt in Saxony ; at Bleyberg in Carinthia; at Idria ; at Salberg, &c. in Sweden ; between the villages of Mandagout and Vigan near Alais in France, it is spread over the soil, which consists of an ochreous earth mingled with quartz and mica,

in long white pieces, which have been taken for
human bones. It is also met with at Kildrummie
and at Portsoy in Scotland.

Mountain wood or *Ligniform asbestus* has somewhat the appearance of wood; its structure is finely
foliated, the foliæ being composed of fine fibres,
which are of a brownish colour. It is opake, somewhat elastic, and floats on water. It is principally
met with in the primitive mountains of the Tyrol,
accompanied with amianthus. It also occurs in
various places in Scotland.

Common asbestus is much heavier than the preceding varieties, being nearly three times the weight
of water. It occurs in masses consisting of fibres
of a dull greenish colour, and pearly lustre. Common asbestus is scarcely flexible. It is of more
frequent occurrence than amianthus : it usually accompanies serpentine; and is met with in Sweden,
Hungary, Dauphiné, the Uralian mountains, at
Porsoy in Scotland, the Isle of Anglesey, and at
the Lizard in Cornwall.

BASALTIC HORNBLENDE.

Basaltic hornblende is usually met with in opake
single crystals, imbedded in basalt or in lava; the
latter sometimes affect the magnetic needle. The
usual colour of this mineral is black; or brownish
black, occasioned by a slight decomposition. The
crystals are six-sided, variously terminated by three
or four planes; but they are sometimes dissimilar

at the two extremities : their primitive form, accord-ing to Haüy, is an oblique rhomboidal prism of 124° 34' and 55° 26' : the crystals have a vitreous lustre, and are hard enough to scratch glass. The specific gravity of this mineral is 3.25 ; and it is composed of 47 per cent. of silex, 26 of alumine, 8 of lime, 2 of magnesia, and 15 of oxide of iron.

Being far less decomposable than basalt, it is sometimes found in fine crystals in the clay result-ing from the decomposition of basaltic rocks. It occurs in Saxony, Bohemia, Italy, Scotland, &c.

HYPERSTHENE.

The Hypersthene is met with either massive, or imbedded in rocks in rhomboidal prisms of about 120° and 60°. Its colour is dark brown, or greenish black ; it has a lamellar structure parallel with the sides of the prism, and when fractured exhibits re-flections which are strongly metallic, and sometimes greenish, sometimes of a copper red colour; it is opake, and yields to the knife. Its specific gravity is 3.38; and it consists of 54.25 of silex, 2.25 of alumine, 1.5 of lime, 14 of magnesia, 24.5 of oxide of iron, and 1 of water.

It usually occurs in serpentine ; and is thus found in Cornwall associated with compact felspar; it is likewise found at the Col de Cervière in the Alps, at Matray in the Tyrol, at Basta in the duchy of Wolfenbuttel, and in Hungary, &c.

It is very nearly allied to the following substance.

SCHILLER SPAR.

Schiller Spar, like the preceding mineral, is always found in serpentine, in which it generally
occurs disseminated. It is of an olive, or bottle
green colour, and when held in certain directions,
has a shining lustre, nearly approaching that of some
of the metals : it is opake and yields to the knife.
A principal difference between the schiller spar and
hypersthene is, that the former fuses, though with
some difficulty, into a black enamel; the latter is
infusible. By one analysis it yields 41 of silex, 3 of
alumine, 1 of lime, 29 of magnesia, 14 of oxide of
iron, and 10 of water.

It is met with in the serpentine of Cornwall and
of Anglesey, and generally speaking, wherever the
hypersthene is found.

AUGITE. PYROXENE.

Augite usually occurs in translucent six-sided
crystals, terminated by dihedral summits; they are
of a blackish-green colour, variously mixed with
brown : it is also met with in angular and rounded
pieces. The form of the primitive crystal is an
oblique rhomboidal prism of 87° 42° and 92° 18′.
It scratches glass with ease. Its specific gravity is
about 3.3 ; and it is composed of 52 of silex, 3.3 of
alumine, 13.2 of lime, 10 of magnesia, 14.6 of oxide
of iron, and 2 of oxide of manganese.

Augite is met with in the productions of volcanoes; but whether it existed in certain rocks, previously to their being subjected to volcanic action, or whether it has been formed in the lavas and scoriaceous matters in which it is found, since their ejection, is matter of uncertainty and dispute. The greater number of mineralogists incline to the former opinion.

It is found in the volcanic countries of Vesuvius, Etna, Stromboli, Auvergne, &c.

It is also said to occur in the basalts of Bohemia, Hungary, Transylvania, Hessia, and in the iron mines of Arendahl, in Norway. The crystals met with in basalt are larger, of a finer green, and more brilliant than those found in lavas.

The coccolite and sahlite are regarded as varieties of augite.

The *coccolite* is of various shades of green, and occurs in little round translucent masses, or in grains of irregular shapes, which are very slightly coherent, but are hard enough to scratch glass : the structure is lamellar, and the lustre vitreous. It consists of 50 per cent. of silex, 1.5 of alumine, 24 of lime, 10 of magnesia, 7 of oxide of iron, and 3 of oxide of manganese.

It is said to have been met with only in primitive countries; in certain veins near Arendahl, in Norway, and Nericia, in Sweden; and in the iron mines of Hellesta and Assebo, in Sudermania.

The *sahlite* occurs in crystals of which the prisms are four or eight-sided, and the summits diedral, and which are of a greenish-grey colour, and scarcely hard enough to scratch glass; they are translucent

c

on the edges. The sahlite is composed of 53 of silex, 3 of alumine, 20 of lime, 19 of magnesia, and 4 of oxide of iron and manganese.

It has been found in the silver mine of Sahla (whence its name) in Westmania, in Sweden, and at Buoen, near Auen, in Norway. It has also been met with in the mountain of Odon-Tchelon, in Siberia, accompanied by mica, beryl, and crystallized phosphorescent carbonate of lime.

PYROPE.

The Pyrope occurs in round or angular grains, of a blood-red colour, which is sometimes clouded with yellow; it never is found crystallized. It is transparent, has a conchoidal fracture, and vitreous lustre, and is hard enough to scratch glass. Its specific gravity is about 3.8, and it is composed of 40 per cent. of silex, 28.5 of alumine, 3.5 of lime, 10 of magnesia, and 16.75 of oxide of iron and manganese. It is sometimes, from its general colour, ranked among garnets; from which it essentially differs in respect of form and composition.

It occurs imbedded in serpentine at Zoeblitz, in Saxony, and in wacke, in Bohemia; but is more common in the latter country in alluvial deposites, accompanied by hyacinths and sapphires. It is met with in the sand of the sea shore at Ely, in Fife-shire; and in Cumberland in clay-stone.

Pliny and Ovid mention a stone by the name of Pyrope, which is supposed to be nearly allied to this mineral.

POTSTONE.

This substance is found massive; such is its structure, that it is sometimes difficult to distinguish it
from massive talc; its colour is greenish-grey, passing into leek-green, with a glistening or pearly
lustre; it is so soft as to yield to the nail, and is
unctuous to the touch, but is not easily broken;
that of Chiavenna consists of about 38 parts of silex,
7 of alumine, 35 of magnesia, 15 of iron, together
with very small portions of lime and fluoric acid.

Potstone is plentifully found at Chiavenna, in the
Valteline; at Coma, in Lombardy; and, generally
speaking, in serpentine countries. Its infusibility,
joined to its softness, and the readiness with which
it is turned by the lathe, have for time immemorial
caused it to be formed into vessels in the Valais and
Grisons. Pliny describes its having been used in
like manner in his time.

SMARAGDITE.

The Smaragdite is of a brilliant green colour, of
a silky or pearly lustre, and transparent at the edges,
or opake: it is scarcely so hard as glass, and yields
to the knife; its specific gravity is 3; and it is composed of 50 of silex, 21 of alumine, 13 of lime, 3
of magnesia; the remainder being oxide of chrome
and oxide of iron.

It is commonly found massive, or disseminated

in rounded masses of the Saussurite, on the banks of the Lake of Geneva : near Turin it occurs at the foot of the mountain Mussinet ; in Corsica imbedded in felspar.

ACTINOLITE.

This mineral is of a pale or of an emerald-green colour, and occurs in single crystals, but more often in masses consisting of diverging hexahedral prisms, which, in the general, are not regularly terminated; they have a shining pearly lustre, and are translucent or transparent; it also occurs in fine fibres, having a silky lustre. Actinolite is hard enough to scratch glass : its specific gravity is about 3.3; and it is composed of 50 per cent. of silex, 0.75 of alumine, 9.75 of lime, 19.25 of magnesia, 11 of oxide of iron, 5 of oxide of chrome, and 3 of water. The fibrous variety is distinguishable from Amianthus by its being extremely brittle.

Actinolite is found only in some of the primitive rocks, and accompanies talc and mica. It is not found in secondary rocks, or in the veins that traverse them.

It occurs in long six-sided prisms imbedded in white talc, at Zillerthal, in the Tyrol, and in Mount St. Gothard; it is also met with near Salzburg, in Saxony; in Norway; in Piedmont, &c.

COLOPHONITE.

This mineral is of a blackish or yellowish brown, or of an orange-red colour; and is, both on the surface and when fractured, of a shining vitreous lustre. It is usually ranked as a variety of garnet, but differs from it in yielding by analysis both magnesia and oxide of titanium, and in being much lighter: its specific gravity is only 2.5; and it is composed of 35 per cent. of silex, 15 of alumine, 2.9 of lime, 6.5 of magnesia, 7.5 of oxide of iron, 4.75 of oxide of manganese, and 0.5 of oxide of titanium It is found near Pitigliano, in Italy

LEUCITE.

The Leucite occurs in crystals, whose planes are 24 equal and similar trapeziums: by mechanical means it may be reduced either to the rhomboidal dodecahedron, or the cube; the latter of which, being the most simple of the two, is considered to be the form of the primitive crystal. The Leucite is generally of a dirty white colour, and is somewhat translucent; it scratches glass with difficulty; its fracture is imperfectly conchoidal, and has mostly a vitreous lustre. It consists of 53.75 of silex, 24.62 of alumine, and 21.35 of potash. Its specific gravity is 2.37.

The Leucite is most commonly found among the productions of volcanoes; that which occurs in lava is mostly opake and earthy, while that found in basalt is vitreous. The lavas of Vesuvius, and ba-salts of Italy and Bohemia abound with this mineral. The road from Rome to Frascati is in many places quite covered with it.

LITHOMARGA.

Lithomarga varies in colour from white to yel-low, red and brown; it is dull, yields to the nail, is unctuous to the touch, and adheres strongly to the tongue; its fracture is mostly earthy.

It is found in masses, somewhat round, in basalts and amygdaloids; and is often met with in veins passing through porphyry, gneiss, serpentine, &c. sometimes it accompanies tin, mercury, and topazes. It seems therefore chiefly to belong to primitive countries; it occurs in France; at Laschitz, in Bo-hemia; at Planitz, near Zwickau, in Saxony, and at Steinmark. That brought from the latter place consists of about 45 parts of silex, 36 of alumine, 3 of iron, 14 of water, and a small portion of potash.

MICA.

Mica mostly occurs crystallized in six-sided plates, or in right rhomboidal prisms of 60° and 120°, which is considered to be the form of its primitive crystal. It is easily divisible, parallel with the terminating planes, into thin laminæ, which are

flexible and very elastic ; this last character serves
at once to distinguish mica from talc, which is not
elastic.

This mineral is of various shades of white, yellow,
green, and brown ;—it yields readily to the knife, but
the edges of the laminæ will scratch glass. The mica
of different countries does not perfectly agree in
the respective proportions of its ingredients ; that
of Muscovy (called *Muscovy Glass*) consists of
about 48 silex, 34 alumine, 9 potash, 4 oxide of
iron, and nearly 1 of oxide of manganese. Its sp.
gr. is about 2.7.

Mica is one of the most abundant mineral substances : it is never found in beds, or in considerable isolated masses, but it enters into the composition of very many rocks, especially the oldest primitive, as granite, gneiss, micaceous schistus, &c. and
is often found filling up their fissures, or crystallized
in the cavities of the veins which traverse them.
Mica is therefore of the most ancient formation ;
but is also met with in the newest crystalline rocks.
It also occurs in sandstones, in schists, and in the
slaty sandstone that accompanies the independent
coal formation. It is sometimes abundant in sands,
and in alluvial deposites very distant from primitive
mountains ; and is said to be very plentiful in certain volcanic products.

According to Haüy, Muscovy Glass, which occurs
in plates of a yard or more in diameter, in veins of
granite and of micaceous-schistus, in some parts of
Russia, may be divided into plates no thicker than
$\frac{1}{300000}$th part of inch. It is used for inclosing

objects for the solar microscope, and instead of glass in the Russian ships of war, as less liable to be broken by the concussion of the air, during the discharge of heavy artillery; an inferior kind, which is found in Pennsylvania, is used there instead of window glass.

MESOTYPE.

The Mesotype is generally of a white, or greyish colour, and is transparent, or translucent; it yields easily to the knife, and becomes electrical by heat. It occurs crystallized in radiated acicular prisms; in filaments; or in globular concretions, composed of stellated fibres. It is one of those substances which are commonly called *Zeolites*. It assumes about 10 varieties in the form of the crystal, the primitive of which is a right prism with square bases. Its specific gravity is 2; and it consists of 49 of silex, 27 of alumine, 17 of soda, and 9.5 of water, according to Simpson; but according to Vauquelin, 50.24 of silex, 29.3 of alumine, 9.46 of lime, and 10 of water.

Mesotype is found in Iceland; Scotland; the Ferroe islands; in Hessia; the Isle of Bourbon, &c.

This mineral is generally considered to be of doubtful origin. It is found in lavas, but principally, if not only, in those that are ancient; and, it is said by some, only in such as have been exposed to the action of water. It is also met with in basalts; as in those of the Giant's Causeway in Ireland; and in those of the Cyclop Islands, and of the Vicentine mountains; the basalt of the two latter is surrounded and covered by the remains of sea animals. The mesotype also occurs in basalt, amygdaloid, and other trap rocks of England and Scotland, and is particularly abundant and beautiful at Talesker, in the Isle of Sky.

The *Natrolite* is composed of the same elementary substances, and very nearly in the same proportions, and is therefore considered to be merely a variety of the mesotype. It is always of a fibrous and radiated structure; and is of a whitish, yellowish, or of a brown colour. In its cavities are found crystals presenting the form of common mesotype, viz. a rectangular prism, with tetrahedral pyramids.

RUBELLITE.

The Rubellite is of a red or violet colour, and occurs crystallized, but the crystals are rarely distinct. It is found in Moravia; in Ceylon; it occurs in a granite mountain in the Uralian chain in Siberia, in a vein composed of felspar, quartz, mica, and common schorl; whence this mineral has been also called *Siberite.* It consists of 42 per cent. of silex, 40 of alumine, 10 of soda, and 7 of oxide of manganese and iron. It is commonly considered to be a variety of tourmaline, from which it differs, in not having either lime or magnesia among its constituent elements, and in being infusible. This mineral is commonly known by the name of *Red Schorl.*

58 *SILEX, Alumine, Potash, Soda, iron, oxygen.*

PUMICE.

Pumice is sometimes found massive; more often it is extremely porous, of a fibrous structure, and harsh to the touch; its colour is grey, tinged with brown or yellow, and it has a shining pearly lustre; it is translucent in the edges, very light, and sometimes so light as to swim on water. It is composed of 77.5 parts of silex, 17.5 of alumine, 1.75 of oxide of iron, and 3 soda and potash.

Pumice is generally believed to be a volcanic product; it sometimes accompanies obsidian; it is said that the vitreous obsidian of Hungary, may, by heat, be changed into a substance perfectly resembling pumice.

It is but sparingly found near Vesuvius, not at all near Etna. It is very abundant in the Lipari islands, which furnish the pumice of commerce. It is met with in Auvergne in France, in Iceland, Teneriffe, &c.

ICHTHYOPHTHALMITE; or FISH-EYE-STONE.

At first sight this mineral resembles the variety of felspar called adularia, but is much softer, being easily cut by the knife; it does not scratch glass. Its general colour is white, which is sometimes tinged with red or green; it has a shining pearly lustre. The form of its primitive crystal is a rectangular parallelopiped, in which it sometimes occurs; as well as nearly in the proportions of the cube, and in flat tables. Its specific gravity is

2.46; and it is composed of 51 parts of silex, 28 of lime, 4 of potash, and 7 of water.

It is met with in the iron mine of Utoe in Sweden ; its gangue is a lamellar carbonate of lime, of a red violet colour; it is accompanied by hornblende and some ores of iron. The massive occurs at Dunvegan in the Isle of Skye.

TALC.

Talc is for the most part either white, apple-green, or yellowish. It occurs in hexagonal laminæ, and massive. It always consists of plates or laminæ, which are easily separated from each other, and are flexible, but not elastic. This last character serves to distinguish this mineral from mica, which is very elastic. Talc is of a shining lustre, is very unctuous to the touch ; yields easily to the nail ; it leaves a white, and somewhat pearly streak, when rubbed on paper. Its specific gravity is 2.77 ; and it consists of 61 of silex, 30.5 of magnesia, 2.75 of potash, 2.5 of oxide of iron, and 0.5 of water.

Crystallized talc, which is mostly white, or of a light green colour, is met with in small quantities in serpentine rocks, with actinolite, carbonated lime, steatite, compact talc, &c. It is found in the mountains of Salzburg and the Tyrol, and is taken to Venice ; whence it has obtained the name of *Venetian Talc.* It occurs also at Briancon ; at Zœblitz in Saxony ; in Silesia, &c.

Massive talc is less flexible and translucent than

c 6

the crystallized; it is principally of an apple-green colour, and is sometimes of a radiated structure. It is met with in considerable beds in mountains of micaceous schistus, gneiss, and serpentine. At Grenier in the Tyrol, it occurs in a species of serpentine, accompanied by actinolite, carbonate of lime, sulphuret of iron, green mica, &c. At Zillerthal, in the Tyrol, it is met with enclosing long prisms of actinolite, and of tourmaline. It occurs also in Austria, Stiria, &c.

Talc is found in Glen Tilt, in Perthshire, in a granular limestone.

Indurated Talc, of a greenish-grey colour, and massive, is met with at the Lizard, in Cornwall, which is a serpentine country.

GREEN EARTH.

This mineral is met with in small masses, or lining the cavities of amygdaloid; and is of a greyish or bluish-green colour, passing into blackish green; it is dull, and yields to the nail; its fracture is generally earthy. It is found wherever amygdaloid occurs; as in Saxony, Bohemia, Monte Baldo near Verona, the Hill of Kinnoul near Perth in Scotland, &c. That of Verona consists of 53 of silex, 2 of magnesia, 10 of potash, 28 of oxide of iron, and 6 of water. When of a good colour it is made some use of by painters.

A substance of a green colour may be observed in little round masses in certain sand stones, as in that of the coast near Folkstone, which is, by some, considered to be a variety of green earth.

SPODUMENE. TRIPHANE.

This rare mineral is of a greenish white colour, of a shining pearly lustre, and translucent. It considerably resembles adularia, but differs essentially from it in respect of mechanical cleavage. Spodumene is divisible into prisms with rhombic bases, having alternate angles of 80° and 100°. It is hard enough to scratch glass, and to give sparks by the steel: its specific gravity is 3.192, and it is composed of 64.4 of silex, 24.4 of alumine, 3 of lime, 5 of potash, and 2.2 of oxide of iron.

It has only been found in the iron mine of Utoe, in Sweden, in a gangue of red felspar, fat quartz, and black mica.

FELSPAR.

Felspath, in German, signifies rock-spar: feldspath, field-spar.

The general form of the crystals of felspar is an oblique prism, having very unequal planes; Haüy notices 21 varieties: the structure is lamellar, and felspar may be cleaved into an oblique angled parallelopiped, which therefore is the primitive form.

The alliance of the crystals with each other is not easily traced, on account of the great difference frequently existing in the size, and consequently in the form, of its secondary planes, as well as on account of its being often in hemitrope or macled crystals ; it is hard enough to scratch glass, but not so hard as quartz, and yields to the knife with some difficulty ; it becomes phosphorescent by friction. There are several varieties of felspar.

Adularia, so called from its having been first met with on one of the heights of St. Gothard, called Adula, is found both massive, and crystallized ; it is of a greenish white colour, but almost limpid, and has a pearly lustre ; its fracture is imperfectly conchoidal. Its specific gravity is 2.54 ; and it consists of 64 per cent. of silex, 20 of alumine, 2 of lime, and 14 of potash. In the veins of mount St. Gothard it occurs in large and well defined crystals in gneiss and micaceous schistus ; and in the mountains near Mont Blanc, in crystals much smaller and less transparent.

The *Moon-stone,* so called from its pale white hue, is considered to be a kind of adularia ; and is brought from the East, particularly from Arabia and Persia.

Common felspar occurs of a whitish, yellowish, reddish or red colour, and either granular, massive, disseminated or crystallized ; it is sometimes opake, sometimes translucent ; its specific gravity is 2.54, and it is composed of 62.83 parts of silex, 17.02 of alumine, 3 of lime, 13 of potash, and 1 of oxide of iron.

Common felspar is the most generally diffused, both as to its local and geological situation, of any other mineral, except quartz and oxide of iron. It is an essential constituent of granite and gneiss, and frequently occurs in micaceous and argillaceous schistus; it forms a large proportion of sienite, and is contained in almost all porphyries, in some very abundantly: it is occasionally, though rarely, found in primitive limestone: it abounds in primitive and secondary traps, and in the greater part of real lavas.

A variety of a beautiful apple *green* colour has been met with only in a hill at the eastern base of the Uralian mountains, near the fortress of Troitzk.

Felspar is occasionally met with, which is more *compact* than the common, but agreeing with it in most respects, except that its structure is less decidedly lamellar, and that its specific gravity is greater; being 2.63.

Lamellar Felspar. Petuntze. Under these names has been described felspar in the first stage of decomposition, but preserving its lamellar character. Its ordinary colour is dirty white, and it sometimes occurs in large masses, enclosing small portions of quartz. It is chiefly employed in giving the enamel to porcelain. The manufactories of France are chiefly supplied from the neighbourhood of Limoges. A slightly saline taste belongs to it, which also is characteristic of the petuntzé of China. The perfectly disintegrated felspar, being usually considered as one of the clays, is noticed with them under the name of *Kaolin*.

Glassy Felspar. Sanidin. This mineral is chiefly found in crystals, sometimes longitudinally striated; it occurs imbedded in porphyry-slate, in Bohemia, at Drachenfels near Born on the Rhine, at Solfatara in Italy, and in Pitchstone in the Isle of Arran. It obtained its name of Glassy, from its vitreous lustre, which sometimes approaches to pearly : it is semi-transparent and translucent, and of a greyish or yellowish white colour. Its specific gravity is 2.57 ; and it is composed of 68 parts of silex, 15 of alumine, 14.5 of potash, and 0.5 of oxide of iron.

Labrador Felspar. The beautiful and varied tints of this mineral, when viewed in particular directions, are well known ; it has the usual characters of felspar, except that its general colour is grey, or dark ash grey ; and that, by the analyses of this mineral, which are not greatly relied on, it appears that potash does not enter into its composition. Its specific gravity is 2.6.

It was first discovered by the Moravian missionaries in the island of St. Paul, on the coast of Labrador ; it has since been found in Ingermannland in Norway ; near the lake Baikal in Siberia ; in granite near St. Petersburg ; also at Memelsgrund in Bohemia, and near Halle in Saxony. It is sometimes accompanied by mica, schorl, and iron pyrites.

SCALY TALC. NACRITE.

This mineral occurs in minute aggregated scales, of a silvery white or greenish colour, and of a glimmering pearly lustre; they are friable, very unctuous to the touch, light, and adhere to the fingers. Scaly talc is composed of 50 per cent of silex, 26 of alumine, 1.5 of lime, 17.5 of potash, 5 of oxide of iron, and a small portion of muriatic acid. Its colour distinguishes it sufficiently from chlorite; it differs from the lepidolite principally in respect of colour, and in being extremely unctuous.

It is chiefly met with in small masses in the cavities of primitive rocks, and in the interstices of crystallized quartz. It occurs at Sylva in Piedmont, near Freyberg in Saxony, and near Meronitz in Bohemia.

PEARLSTONE.

Pearlstone occurs in large coarse angular concretions, including smaller round concretions, composed of very thin lamellæ. The surface is smooth and shining, with a lustre remarkably resembling that of pearl. The colour of the mass is grey, greyish black, brown, redish or blackish. It is fragile, translucent on the edges, and scarcely hard enough to scratch glass. Its specific gravity is 2.34; that of Hungary is composed of silex 75.25, alumine 12, lime 4.5, potash 4.50, oxide of iron 1.6, and water 4.5. It almost always gives out an argillaceous smell when breathed on. Some of the varieties are said to bear a striking resemblance to pumice.

At Tokay in Hungary, it is found enclosing round masses of black vitreous obsidian, and is intermixed with the debris of granite, gneiss, and porphyry, and alternating in beds with the latter. A variety met with at Cenapecuaro in Mexico, is hard enough to scratch glass; another found at Cap de Gate in Spain, of a greenish or bluish colour, does not give out the argillaceous odour. Pearlstone is also met with at Sandy Brae, in the island of Egg, one of the Hebrides.

AGALMATOLITE.

This mineral obtained the French and German names of Pierre de Lard and Bildstein, from the resemblance of some of its varieties to Lard; and Brongniart has given it that of steatite pagodite, from its being always brought from China in the form of little grotesque figures and chimney ornaments; but all the analyses of it, distinguish it sufficiently from steatite, which is always in part constituted of magnesia. The agalmatolite is also found at Nagyag in Transylvania. It consists of 56 of silex, 29 of alumine, 2 of lime, 7 of potash, 1 of oxide of iron, and 5 of water. In the varieties of the Chinese, analyzed by Klaproth, no indication of potash was found, and one of them was without lime.

LEPIDOLITE.

The Lepidolite is of a pearl grey, rose red, or of a lilac red, or purple colour, whence it has also been called the *Lilalite.* It consists of an assemblage of small flexible scales, which are translucent: the mass has a pearly or silvery lustre, yields to the nail, and is somewhat unctuous to the touch. Its specific gravity is 2.85: that of Moravia consists of 54 per cent. of silex, 20 of alumine, 4 of fluate of lime, 18 of potash, 4 of oxide of manganese, and 1 of iron.

It was first discovered on the mountain Gradisko, near Rozena, in Moravia, of a pale rose colour and pearly lustre; it occurred also in a thin bed in gneiss, accompanied by quartz, mica, schorl, &c. It has since been met with in Sweden in a quartzose rock; in France, near Limoges, in a vein of quartz, passing through granite, enclosing large beryls; at Campoin, in the island of Elba, of a rose colour, in a rock composed of quartz and felspar.

OBSIDIAN.

Common obsidian is of a greenish or brownish black, or of a smoke brown colour, with a shining vitreous lustre; its fracture is conchoidal; some varieties are translucent, others nearly opake, and it is hard enough to scratch glass: its specific gravity is about 2.35. That of Hecla yields by analysis 78 of silex, 10 of alumine, 2 of lime, 6 of potash, 1 of oxide of iron, and 1 of manganese. Potash and lime do not enter into the composition of all the varieties. It occasionally very much resembles common glass.

The origin of obsidian has been very warmly contested; it is most common in the neighbourhood of Volcanoes, and has been considered as a vitrified lava; whence it has obtained the familiar name of *Volcanic glass.* It occurs in beds, masses, and in small isolated pieces.

Fragments of blackish obsidian are met with, not only at the foot of Hecla, but in almost every part of Iceland: It is also found in the Lipari islands; some varieties enclose felspar. In Peru it is met with in parallel beds of a greenish black, and greyish colour; the latter enclosing opake, spherical masses, of a slate colour, composed of diverging fibres. In New Spain, some obsidians, which have been long exposed to the air, are covered by a white opake enamel.

Obsidian, of a greenish black colour, constitutes the greater part of the mountain della Castagna, in the island of Lipari; it encloses small crystals of felspar; and near the peak of Teneriffe obsidian appears in the form of considerable currents, (like lava) presenting some fibrous appearances, denoting its passage into Pumice.

A variety of a silky and chatoyant lustre is also found in New Spain.

Obsidian in the form of little grains of the size of peas, of a pearly white, and consisting of very thin concentric layers; together with fragments of these; also vitreous globes of the size of a nut, and others like enamel, traversed by red and black veins; forming altogether a species of vitreous sand, is found at Marikan in the Gulph of Kamschatka; and is thence termed the *Marekanite.*

In the island of Ponce, obsidian is met with, enclosing yellow mica, and white vitreous grains, which appear to be semi-vitrified felspar.

Obsidian is in some places traversed by veins of stony or earthy matter of various kinds; thin beds of which also occur between beds of obsidian. In the Madona mountain, in the island of Ponce, the beds are nearly vertical. In Hungary, obsidian occurs, intermingled with the debris of decomposed granite, gneiss and porphyry; and even alternates with beds of the latter. These circumstances have induced some mineralogists to doubt the igneous origin of obsidian; but their strongest arguments are—the violent intumescence which it undergoes when subjected to heat, which causes it to melt into a glass, and the quantity of aqueous vapour disengaged during the process. Humboldt suspects this to be one of the causes of the violent earthquakes so often felt in the Cordilleras of the Andes.

But it is agreed universally, that whenever obsidian is found, there exist indications of volcanic agency in the neighbouring country.

In Europe, obsidian has been fashioned into reflectors for telescopes; in Mexico and Peru, it was made into looking glasses and knives.

HAUYNE. LATIALITE.

The Haüyne is usually found massive, but, in one instance, has been observed in extremely brilliant crystals, but so minute, and crossing each other, in so many directions, that it was impossible to discover their form. When this mineral is opake, it is of an indigo blue colour; when translucent, bluish green. It is somewhat harder than quartz, is very brittle, and its fracture is uneven, and considerably splendent. Its specific gravity is about 3.2; and it consists of 30 per cent. of silex, 15 of alumine, 20.5 of sulphate of lime, 5 of lime, 11 of potash, 1 of oxide of iron, 17.5 water, sulphuretted hydrogen and loss. In some of its external characters and in its chemical composition, it bears considerable analogy to Lapis Lazuli.

It occurs massive in Italy, in the neighbourhood of Nemi, Albano, and Frascati, accompanied by mica, and green pyroxène; and near Vesuvius, its gangue consists of the fragments of rocks ejected by volcanic eruptions, and it is accompanied by idocrase, augite, mica, and meionite.

Haüy seems to be of opinion that the mineral, heretofore termed *Blue Spinelle*, which occurs in the form of a rhomboidal dodecahedron, in the productions of volcanoes at Andernach, on the banks of the Rhine, ought to be considered as a variety of this mineral; as well as the *sapphirin*, which occurs in the granular form on the banks of the lake of Laach, in a rock principally composed of grains, and of small crystals, of vitreous felspar.

ANALCIME.   CUB C ZEOLITE.

The Analcime is usually met with in round or ra‑ diated masses, or in cubic crystals, either perfect, or having each of the solid angles replaced by three planes ; or in the trapezoidal dodecahedron, which is a variety of the cube ; the lustre is shining, and between pearly and vitreous. The colour of the analcime is white, yellowish, reddish, or deep red ; it is hard enough to scratch glass, and is mostly transparent or translucent, occasionally opake ; it becomes electric by rubbing. Its specific gravity is below 3. It consists of 58 of silex, 18 of alu‑ mine, 2 of lime, 10 of soda, and 8.5 of water.

The analcime is sometimes confounded with stilbite, but amongst their distinctive characters, the superior pearly lustre of the stilbite, is that by which they are mostly readily distinguished.

According to Brongniart, this mineral has been met with only among the products of volcanoes ; as in the lavas of Etna : according to Jameson, the cubic zeolite is met with lining the cavities of amygdaloid, basalt &c. : and occurs in Staffa, and near Talysker, in the island of Skye : it is found also in the Hartz, Bohemia, &c. ; in Iceland and the Ferroe islands. At Oberstein, it occurs in the cavities of geodes.

A variety from Somma, called the *Sarcolite,* from

its being of a flesh red colour, is met with in cubes, having each solid angle replaced by planes.

LAVA.

Lava is externally yellowish or greenish grey, greyish black, or greenish black, and is internally spotted reddish, yellowish brown, or grey; sometimes, when sulphureous vapours have acted much upon it, it is yellowish or sulphur yellow. It is vesicular and knotty; the vesicles are empty; sometimes it is porous. Its fracture is imperfectly conchoidal; internally its lustre is glistening or shining. It is opake, translucent on the edges, brittle, mostly attracts strongly the magnetic needle, and, it is somewhat remarkable, is easily fused into a black glass. The compact lava of Calabria yields, by 'analysis, about 51 of silex, 19 of alumine, 10 of lime, 4 of soda, 14 of iron, and 1 of water.

Lava usually encloses crystals of augite, hornblende, felspar, and leucite; which sometimes have no appearance of being altered by heat.

The above description, generally speaking, belongs to those substances, which, by common consent, are true lavas; the products of Etna, Vesuvius, Hecla, and other Volcanoes. But there are many substances, considered by some mineralogists as lavas, which, by others are not allowed to be of volcanic origin. Karsten enumerates nine species of lava; and Haüy six, which are again subdivided;

amongst them are pearlstone and obsidian. Werner notices only two, one of which he calls *Slag Lava,* the other *Foam Lava.* The slag lava is above described ; foam lava is of a greenish grey colour, approaching to greenish black ; it is light, brittle, and often crumbling ; and has often been confounded with pumice.

PITCHSTONE.

The colours of this mineral, which obtained the name of Pitchstone, from the resemblance which some of its varieties bear to pitch, are very various ; it is met with in shades of grey, blue, green, yellow, red, brown, and black ; but its colours are not lively : it has a glistening resino-vitreous lustre. It occurs generally in distinct masses or considerable beds, and has an imperfect conchoidal fracture, which in some varieties is the chief characteristic distinction between pitchstone and obsidian ; and it is not unfrequently confounded with hornstone and semi-opal. It is almost always opake, or only translucent on the edges, and is hard enough to scratch glass. The specific gravity of that of Meissen in Saxony, is 2.32 or 2.64. Pitchstone is composed of 73 per cent. of silex, 14.5 of alumine, 1 of lime, 1.75 of soda, 1 of oxide of iron, 0.1 of oxide of manganese, and 8.5 of water.

The pitchstone of which the analysis is given, is of a yellowish grey colour, and alternates, in the mountain of Gersebach between Meissen and Freyberg, with a porphyry, having a base of petrosilex, which alternates with gneiss, and is traversed by metalliferous veins. Pitchstone is found in veins traversing granite, near Newry, in the county of Down, in Ireland. In these instances the pitchstone, it seems reasonable to conclude, must be of the same origin as the rocks in which it is imbedded. Mineralogists are not agreed in opinion respecting that of pitchstone in general. Those of Planitz in Saxony, and of Cantal in France, are considered to be of volcanic origin. Pitchstone is met with in Dumfrieshire in Scotland, and in several of the Scottish islands.

CLINKSTONE.

The clinkstone is always found massive, and when struck with a hammer, gives a ringing metallic sound; whence its name. It is of a dark greenish, yellowish, or ash grey colour: its fracture in one direction is slaty, and it is hard, brittle, and commonly translucent on the edges. Its specific gravity is 2.57; and it consists of silex 57.25, alumine 25,50, lime 2.75, soda 8.1, oxide of iron 3.25, oxide of manganese 0.25, and 3 of water.

The clinkstone is usually columnar, and generally rests upon basalt. It occurs near Zittau in Upper Lusace; in the Bohemian Mittelgebirge; in South America; in the island of Lamlash in the firth of Clyde; the isles of Mull and Arran; the Ochil and Pentland hills in Scotland; the Breidden hills in Montgomeryshire, and in the Dirris mountain in the county of Antrim in Ireland.

SODALITE.

This rare mineral has only been found associated with sahlite, augite, hornblende, and garnet, in Greenland. Its colour is light green, or bluish green, and it occurs massive, but more often crystallized in rhomboidal dodecahedrons. It is translucent, and yields with difficulty to the knife. Its specific gravity is about 2.37; and according to the analysis of Thompson, it is composed of 38.42 of silex, 27.48 of alumine, 2.70 of lime, 23.5 of soda, 3 of muriatic acid, 1 of oxide of iron, and 2.1 of volatile matter.

CHABASIE.

This mineral is only met with in crystals very nearly approaching the cube, having the edges and sometimes the angles, replaced by planes; but only three varieties of form have been noticed by Haüy, who considers their primitive to be an obtuse rhomboid of 93° 48′ and 86° 12′.

The colour of the Chabasie is white or greyish, sometimes pale red superficially; it is transparent or translucent, and scarcely hard enough to scratch glass. Its specific gravity is about 2.7; and it consists of 43.33 of silex, 26.6 of alumine, 3.34 of lime, 9.34 of potash and soda, and 21 of water.

The Chabasie is met with in the fissures or
cavities of some basaltic rocks, or within geodes of
quartz or agate which are disseminated in rocks. It
is thus found in the quarries of Alteberg, near
Oberstein in Saxony. It is also said to occur in the
lavas of the Isle of Ferroe; at Talisker in the
Isle of Skye; at Glen Farg in Perthshire, and at
Portrush in the North of Ireland.

FETTSTEIN.

The Fettstein has been found only in Norway;
it occurs massive, and of a darkish green, bluish
grey, or flesh red colour; with natural joints parallel to the faces of a right rhomboidal prism; it
is translucent, and scratches glass. Its specific
gravity is 2.6; and, according to Vauquelin it is
composed of 44 of silex, 34 of alumine, 0.12 of
lime, 16.5 of potash and soda, and 4 of oxide of
iron.

The Fettstein is by some mineralogists supposed
to bear considerable affinity to some varieties of
felspar. It has a slight chatoyant lustre when held
in particular directions, like that of Labradore felspar. Both soda and potash enter into the composition of Fettstein, the former predominating; the
latter only is found in felspar.

SCAPOLITE.

The Scapolite is usually met with in prisms of four or eight sides, either terminated by planes or by tetrahedral pyramids, and aggregated laterally. Their colours are grey or yellowish, sometimes with a pearly lustre; or an almost metallic grey; sometimes deep red and opake; occasionally apple-green. The crystals possessing a pearly lustre will scratch glass; but when dull, with an appearance like that of efflorescing, they are tender and even friable. The scapolite is composed of 45 of silex, 33 of alumine, 17.6 of lime, 0.5 of potash, 1.5 of soda, 1 of oxide of iron and manganese; but the efflorescing variety differs, in including some magnesia, and in being without potash.

The Scapolite has hitherto only been met with in the iron mine of Langloe, at Arendahl in Norway; its crystals appear variously grouped, and accompanied by brown mica, quartz, garnet, epidote, carbonate of lime, &c.

JADE.

The general characters of Jade, of which there are three varieties, are, that it always occurs massive, of various shades of green and whitish green, with a greasy lustre; it is unctuous to the touch, harder than quartz, and very tough.

Common Jade is of a leek green colour, passing

into greenish white, semi-transparent, extremely tough, with a glimmering lustre and broad splintery fracture. Its specific gravity is 2.95 ; and it is composed of 53.75 of silex, 1.5 of alumine, 12.75 of lime, 8.5 of potash, 10.75 of soda, 5 of oxide of iron, 2 of oxide of manganese, and 2.25 of water.

Of the geological history of common Jade, nothing is known. It is found in Switzerland, Piedmont and Tyrol ; China and India. It is regarded in the latter countries as a specific for the nephritic cholic, and is fashioned into forms of great delicacy. The Hindoos and Chinese form it into talismans and idols ; the Turks into swords and dagger handles.

The *Axestone*, or *Beilstein*, differs from common jade in having a slaty structure, and in being less transparent and less tough. In America, it is found in the banks of the river Amazons ; whence it obtained the name of the *Amazonian stone*. It is also met with in Corsica, Switzerland and Saxony ; and in New Zealand and other islands in the Pacific ocean, where it is made into hatchets, tomahawks, and other instruments ; whence its common name.

The *Saussurite*, or *Tough Felspar*, is greener than the preceding varieties, and at least as hard and as tough as common Jade : according to Saussure, it consists of 44 parts of silex, 30 of alumine, 4 of lime, 0.25 of potash, 6 of soda, 12.5 of oxide of iron, and 0.5 of oxide of manganese.

It was first found by Saussure, whence its name, in rounded masses on the edge of the lake of Geneva, and afterwards near Turin, in the mountain Mussinet, which is principally composed of serpentine, which encloses hydrophane. It has since been met with in Corsica; in sand, in the neighbourhood of Potsdam, and near Affschaffenberg.

The Soapstone is found massive, and nearly white or of a grey colour, sometimes with a tinge of yellow, and mottled with green or purple; it is translucent on the edges. Its fracture is somewhat splintery; it yields to the nail; from its general aspect and unctuous feel, its name has been derived.

It is met with in a vein in serpentine at the Lizard point in Cornwall, where it may sometimes be found with the appearance of passing into asbestus, which occurs in veins in the serpentine. It is much used in the manufactory of porcelain. It also occurs near the Cheesering, at St. Cleer, in Cornwall.

The soapstone of Cornwall consists of 45 per cent. of silex, 9.25 of alumine, 24.75 of magnesia, 0.75 of potash, and 1 of oxide of iron. It is commonly supposed to be a variety of steatite, but is much softer. In the composition of the latter, no alumine has been detected.

CHLORITE.

Chlorite is composed of very minute plates intersecting each other in various ways, giving to the mass a granular or earth structure: it also occurs

D 4

crystallized in flat six sided crystals, which are readily divisible into thin laminæ. It is usually of a
dark green, sometimes of a yellowish green, with a
shining lustre; it is opake, yields to the nail, is
somewhat unctuous; and when massive, gives out
an earthy smell when breathed on.

Common Chlorite is usually found massive and
somewhat solid; its specific gravity is 2.56.

Common chlorite is not found in very considerable masses; but chiefly in the veins and cavities
of primitive rocks; sometimes it is enclosed in crystals of quartz, chalcedony, felspar, axinite, &c. in
so large a proportion as to impart a colour to them.
It frequently accompanies the oxide of tin and mispickel in the veins of Cornwall; and occasionally,
though rarely, yellow copper. It is met with in
most chains of primitive mountains.

When the structure of chlorite is slaty, it is
termed *Chlorite slate*; its specific gravity is greater
than that of the preceding variety, being 3.03. Its
ordinary colour is blackish brown. It is met with
in beds in primitive mountains, enclosing crystals
of quartz, octohedral magnetic iron ore, garnets, &c.
and is found in Corsica, at Fahlun in Sweden, in
Norway, &c.; and in Perthshire.

Scaly Chlorite is of a dark green colour, and is
composed of small glimmering particles having a
pearly lustre; it is somewhat unctuous to the touch;
is friable, or loose; and greatly resembles green
earth. It is very light and consists of 26 of silex,
18.5 of alumine, 8 of magnesia, 2 of muriate of soda
or of potash, and 43 per cent. of oxide of iron.

It mostly occurs in the veins of primitive mountains, principally in clay slate, mixed with quartz, common chlorite, calcareous spar, and micaceous iron ore : it is also met with in granular limestone, and in primitive sandstone. It is found in Saxony, Switzerland, Savoy, Sweden, Hungary, and North Wales.

SCHORL.

Schorl is found massive, disseminated, and crystallized ; the common form of the crystals is a prism mostly striated longitudinally and deeply, and terminated at each end by 3 planes ; but the crystals are sometimes very minute, closely aggregated, and divergent. This substance is black, brittle, opake, and has a glistening lustre. Its specific gravity is about 3.2. and it is composed of about 38 parts of silex, 34 of alumine, 1 of magnesia, 6 of potash, 21 of oxide of iron, and a trace of manganese.

Schorl, except by the Wernerian schorl, is arranged among tourmalines, from which it differs in respect of analysis, transparency and colour. The latter mostly occur imbedded in single crystals ; the former is mostly aggregated, and occurs in beds.

It is found in primitive rocks ; chiefly in quartz and granite ; more rarely in gneiss and micaceous schistus ; and is frequently met with in tin veins.

Schorl was first found near the village of Schorlaw in Saxony, whence its name. It is also met with in Bohemia, Bavaria, Switzerland, Spain and Hungary ; and at Portsoy in Scotland, and beneath the Logan

Rock, and at various places near the Lands' End in Cornwall.

CLAY-SLATE. ARGILLACEOUS SCHISTUS.

The prevailing colour of clay-slate is grey of various shades; it is also bluish or blue; and sometimes greenish, passing into blackish green. Its structure is slaty, and it has a glistening lustre, sometimes approaching to pearly; it is opake, and yields to the knife, but varies in hardness, and some varieties are somewhat unctuous to the touch. Its specific gravity is about 2.7; and it is composed of 48 per cent. of silex, 23.5 of alumine, 1.6 of magnesia, 11.3 of oxide of iron, 0.5 of oxide of manganese, 4.7 of potash, 0.3 of carbon, and 7.6 of water.

Clay-slate occurs in vast strata in primitive mountains, and sometimes in veins. It is very universally distributed. In Britain; it is met with in Scotland and the Scottish isles, in the northern parts of England, and plentifully in Cornwall, being the *Killas* of the miner. The principal part of the numerous copper and tin mines of that county are situated in clay-slate; which in most countries abounds in mineral veins.

Some varieties which readily split into thin plates are used for the roofing of houses; another is used for writing on; another as pencils; and some varieties as whetstones.

GABRONITE.

The Gabronite occurs massive, and is of a bluish or greenish grey colour; its fracture is lamellar, it is translucent on the edges, and hard enough to scratch glass, though not to give fire with the steel. Its specific gravity is nearly 3; and it is composed of 54 per cent. of silex, 24 of alumine, 1.5 of magnesia, 17.25 of potash and soda, 1.25 of the oxides of iron and manganese, and 2 of water.

The Gabronite has only been found in Norway. The bluish variety, near Arendahl, with hornblend; the greenish, at Fredericksvarn, disseminated in a large grained sienite.

FULLER'S EARTH.

Fuller's earth occurs massive, and is usually of a greenish brown colour, sometimes nearly of the colour of slate; it is dull, possesses an earthy fracture, and yields to, and receives a polish from, the nail: in water it becomes semi-transparent, and falls into a pulpy impalpable powder. The English Fuller's earth is composed of 53 of silex, 10 of alumine, 0.5 of lime, 1.25 of magnesia, 9.5 of oxide of iron, 1 of muriate of soda, and 24 of water.

At Nutfield, near Riegate, in Surry, it occurs in regular beds near the summit of a hill of considerable elevation, between beds of ferruginous sand or sandstone containing fossil wood, cornu ammonis, impressions of the nautilus and other sea-shells. There are two distinct beds of Fuller's earth; the upper, of a greenish clay colour and 5 feet in thicknesss, rests upon the other, which is of a light slate blue, and 11 feet thick; in these beds, but mostly in the latter, are found considerable masses of sulphate of barytes, sometimes exhibiting regular crystallizations, the interstices of which are occasionally filled up by compact quartz.

Fuller's earth is also found at Deptling, near Maidstone in Kent, and at Aspley, near Woburn in Bedfordshire, under nearly the same circumstances as at Nutfield. At Old Down near Bath, it occurs mixed with shells, forming a bed between the upper and under oolite; and near Nottingham in lumps in the red marl.

It is found near Rosswein in Saxony, under very different circumstances to that of England. It occurs among primitive rocks, and is supposed to originate in the decomposition of green-stone-slate, beneath which it lies.

Fuller's earth was formerly much used in the fulling of cloth (whence its name), and was forbidden to be exported under severe penalties: soap is now generally substituted.

BASALT.

Basalt is of a greyish black colour, and when polished, of a bluish aspect. It is not easily broken; its fracture is dull, but fine grained. Some

varieties strike fire by the steel, others may be
scratched by the knife. It has a tendency to
form 6 sided irregular prisms or pillars; of which
the Island of Staffa is entirely composed. The
Giant's causeway, on the coast of Antrim in Ireland, is a huge pavement of strait pillars of Basalt,
running to an unknown distance into the sea; and
the promontory of Fair-head, a little further north,
exhibits a continued range, about a mile in length,
of columns 250 feet high, and from 10 to 20 in diameter, being the largest yet known. When exposed
to weather, basalt crumbles down into a fine black
mould, which constitutes a very fertile soil. It is
to this rock that some of the richest parts of Scotland owe their fertility.

The basalt of Saxony is composed of 44.5 per cent,
of silex, 16.75 of alumine, 9.5 of lime, 2.25 of magnesia, 2.6 of soda, 20 of oxide of iron, 0.12 of oxide
of manganese, and 2 of water.

Basalt is found under very different circumstances: it occurs filling up veins and fissures in many
primitive and secondary mountains; sometimes forms
beds or strata on their summits; and not unfrequently, it traverses coal-formations, in a direction nearly
perpendicular to the beds of coal, which it seems to
have the effect of dislocating.

Geologists are divided in opinion respecting the
origin of Basalt. Werner supposes it to have been
deposited, like other minerals, by water which
covered the whole earth. Dolomieu conceived basalt
to be lava, and all basaltic mountains to be the remains of extinct volcanoes. Dr. Hutton and Pro-

fessor Playfair conceive it to have been fused by a central fire of the earth, while at the bottom of the sea, and to have been raised up by some natural agent, in common with all other mountains.

With a view to determine the correctness of these opinions, Daubuisson examined the basalt of Saxony, which chiefly lies in the Erzgebürge, or metalliferous mountains; a chain separating Bohemia from Saxony, of about 120 miles long, and 3600 feet high above the level of the sea. The lower rock is granite, which is covered, or rather wrapt round, by beds of gneiss, mica-slate, and clay-slate, lying above each other in that order. In these beds are situated the great mines of Saxony. In a chain of rocks of serpentine and quartz, are found beds of limestone, of coal, &c. The whole of the eastern part of the chain is covered on one side by a huge bed of porphyry, and on the other by a bed of sandstone of equal magnitude.

Basalt forms the summits of about 20 mountains of this chain, under various forms, as of tables, cones, or domes: the mountains are connected by their sides; the basaltic top alone remaining separate.

In several instances, between the basalt and the body of the mountains, he found beds of sand, gravel, and clay: in others, the basalt rests on sandstone; in others, on porphyry; in one, on mica slate; in three, on granite; and in one, on gneiss.

After a complete investigation of these mountains, Daubuisson is of opinion that there is no analogy between them and volcanic mountains. They are regularly stratified, which is never the case in vol-

canic mountains; no trace of a crater can be per-
ceived; nor any thing decidedly volcanic. Besides,
basalt, wherever found, is always composed of the
same constituents; lava varies considerably. The
substances contained in basalt, as felspar, mica, &c.
retain their crystalline characters without exhibiting the slightest traces of the action of fire, though
the felspar is more fusible than the basalt itself.
Basalt contains 20 per cent. of iron, and there
is no rock which could furnish such a proportion.
It contains 5 per cent. of water of composition,
which is never found in lava. It is found lying immediately under or over coal, which is in no degree
altered in its nature; and Dolomieu has described
no less than 20 beds of basalt alternating with as
many beds of limestone containing marine shells.
For these and many other conclusive reasons,
Daubuisson is decidedly of opinion that the Saxon
basalt is altogether of aqueous origin.

He was afterwards induced by some zealous ad-
vocates for the igneous origin of basalt, to explore
the basalt country of Auvergne in France. The
base of this country he found to be granite;
which, in the western part, is covered with gneiss
and micaceous schistus, containing metalliferous
veins. Limestone and coal also appear in other
districts. The chain of the Puys extends above 20
miles: most of them are detached; their form is a
truncated cone; and on their summits there are
cup-like depressions, in some instances 200 feet deep.
Their general elevation is from 900 to 1300 feet

above the plain ; the central and highest, the Puy de Dome being near 2000 feet.

The substances chiefly composing these hills, are, scoriæ, lava, and other decided volcanic matter.

In one instance he traced the appearance of a stream like that of lava, 200 feet broad; which afterwards divided ; the soil it affords is unfruitful.

Its characters differ in some respects from common basalt; the felspar has a vitreous aspect, and the quartz is altered by heat.

Many other circumstances also contributed to induce the full belief that the basalt of Auvergne is of igneous origin.

There seems therefore sufficient grounds for concluding that, as the basalt of Saxony is altogether of aqueous origin, and that of Auvergne of igneous origin, these two rocks ought not, however, in point of composition and aspect, they may resemble each other, to receive the common name of basalt. This seems to be one of the numerous causes of confusion in geological nomenclature.

In Scotland, basalt is included, together with many other rocks of very different natures, in the vague, but comprehensive term *Whin-stone.*

*The Eight following substances have not been ana-
lyzed, but are in most mineralogical arrangements associated with those, of which the principal ingredient is Silex.*

HORNSTONE.

This substance occurs in nodules and massive,
with a splintery fracture, and is translucent, passing
into opake ; it is scarcely so hard as quartz, and is
infusible. Its general colour is grey, which is tinged
blue, green, brown, red or yellow.

Hornstone is described as occurring in round
masses in limestone, as in Bavaria ; and in beds in
limestone on the banks of the Menai in Caernarvonshire ; and sometimes as forming the basis of porphyry ; as in Sweden, at Dannemera and Garben-
burg, and also in the Shetland isles.   Wood, petrified by Hornstone, thence termed *Woodstone* is
met with in ferruginous sand near Woburn in Bedfordshire, and near Nutfield in Surry.   I have met
with Hornstone in Pednandrae Mine in Cornwall,
passing into *Chert*, which is considered to be allied
to it ; the fracture of Chert is flat conchoidal, and it
generally has a waxy or greasy lustre, and is translucent on the edges.   Its general colour is grey.

There is a considerable bed of Chert near the
summit of the Cliff at the Western lines in the
Isle of Wight, resting, I believe on sandstone,
which having given way by exposure, the fall of the
Chert has been the principal occasion of the now

beautiful ruin beneath. Chert also occurs in some parts of Devonshire and Dorsetshire, and is employ-ed for repairing the roads. It also occurs resting upon the ferruginous sand or sandstone of Leith Hill in Surry; near the summit of which, the sand, like that beneath the Chert in the Isle of Wight, contains organic remains which are denominated Alcyonia.

CHIASTOLITE.

This mineral has only been met with crystallized in long slender rhomboidal prisms, composed of two distinct substances. The exterior is greyish white or reddish, the interior is black or bluish black, and its sides are perfectly parallel with those of the exterior substance, which in some specimens is so thin as to form a mere coating. From each of the angles of the interior prism, there often proceeds a black line which sometimes reaches the correspond-ing angles of the coating, but is sometimes terminated by a black rhomboidal prism; so that the Chiastolite occasionally consists of 5 black rhomboidal prisms, communicating by black threads, and as it were imbedded in a greyish white or reddish substance, which has a lamellar structure, is translucent, and hard enough to scratch glass.

This substance seems only to have been discovered imbedded in argillaceous or micaceous schistus. In the former, it occurs in the Wolf-crag near Keswick, and on the summit of Skiddaw, in Cumberland; and also at St. Jaques de Compostella in Spain. In the latter, it is met with in the Sierra del Marao in Portugal. The Chiastolite is also found in Brittany in France, in the valley of Barège in the Pyrennees, and at Aghavanagh, and Baltinglass-hill, in the county of Wicklow in Ireland.

SPINELLANE.

The Spinellane has only been found on the borders of the lake of Laach, in a rock composed of grains and small crystals of glassy felspar, quartz, hornblende, black mica, and magnetic iron ore. It occurs in small rhomboidal dodecahedrons of a dark brown colour, and is so hard as to scratch glass.

MELILITE.

This rare mineral has been met with chiefly in small rectangular parallelopipeds; occasionally in rectangular octohedrons. Internally the crystals are of a honey yellow or orange colour; externally they are usually coated by oxide of iron of a yellowish brown colour; they give sparks by the steel. The Melilite has only been found at Capo di Bove near Rome, in the fissures of a compact black lava.

Wacke is of various shades of greenish and yellowish grey, and occurs either solid or cellular:
when the cells are hollow, or filled by some other
substance, as quartz, chalcedony, or carbonate of
lime, the compound is denominated *Amygdaloid.*
Wacke has an earthy fracture, is opake, and generally yields easily to the knife.

According to Werner it occurs in beds, which
generally lie under basalt, and above clay, as at
Fichtelberg, and Marienberg, in the hills of Schne-
genberg; and frequently contains imbedded crystals
of mica, and basaltic hornblende. It also occurs in
veins. At Joachimsthal in Bohemia it encloses
petrified wood, native bismuth, and fragments of
certain primitive rocks; and at Kallennordheim in
Franconia, fossil bones: it is found also at West-
manland in Sweden, and in Iceland.

The *Amygdaloid* or *Toadstone* of Derbyshire is
considered by Werner to be a variety of transition
Trap.

Iron Clay, the *Eisenthon* of Werner, is considered by some as a variety of Wacke; their general
characters are the same; but the colour of the
former is reddish brown; they both occasionally
constitute the basis of amygdaloid.

SHALE. SLATE-CLAY.

Shale occurs only massive; its general colour is grey, which sometimes is bluish, yellowish or blackish; in one direction its structure is slaty, in the other, earthy; it usually adheres a little to the tongue and yields to the nail, and is opake, meagre to the touch, and dull, except from casually imbedded mica, which sometimes imparts a glimmering lustre: its specific gravity is about 2.6.

Shale has the usual characters of clays, by becoming plastic in water; it disintegrates on exposure to air. A variety found at Menil-montant near Paris, enclosing the menilite, yielded 66 per cent. of silex: it adheres strongly to the tongue.

It is found in beds and strata in schist; in alluvial deposites; and resting upon, as well as interposed between, beds of coal, which it invariably accompanies. It often contains impressions of reeds and of ferns; and I am informed by my friend L. W. Dillwyn, well known for his curious botanical researches, that he has never discovered a single impression of fern in shale, perfect as these impressions usually are, exhibiting its well known appearance of fructification.

A variety of shale usually accompanies coal, and is sometimes intermixed with it, which, from its black colour and bituminous quality, is termed *black bituminous shale.* It occurs in every independent coal-formation. Its structure is slaty: when subjected to the flame of a candle, it blazes; in the fire it crackles, emits a black smoke and bituminous odour, loses a considerable portion of its weight, and is converted into a whitish or reddish flaky ash.

Another variety of Bituminous shale, of a *brown* colour, is met with at Kimmeridge in Hampshire, which from its giving out a bituminous odour when placed in the flame of a candle, or in the fire, is termed *Kimmeridge Coal.* By exposure to a considerable heat, the bituminous part is consumed, and it is reduced to a grey earthy ash.

FLINTY SLATE.   INDURATED SLATE.
SILICEOUS SCHISTUS.

Of Indurated Slate, there are two or three varieties.
Common indurated slate. This substance is of about the same hardness as quartz, which commonly traverses it in small white veins. Its colour is very various; grey, bluish grey, and red; its structure is somewhat slaty, and it is translucent on the edges.

A specimen analyzed by Weiglib yielded 75 per cent. of silex, the remainder being lime, magnesia, and oxide of iron.

It is chiefly found in beds in transition mountains, and occurs in Saxony, the Hartz, in the Lead hills and other places in the South of Scotland. At Saaska in the Bannat, and in Greece, it occurs in large masses in transition limestone.

Lydian Stone, or *Basanite*, is of a black or greyish black colour, and is always found massive, never with a slaty structure; it is often traversed by veins of quartz; it is opake, less hard than the foregoing variety, and its fracture is flat conchoidal.

It occurs in similar formations and repositories with common indurated slate; and is found near Prague and Carlsbad in Bohemia; near Freyberg in Saxony; and in the Moorfoot and Pentland Hills near Edinburgh. It was first brought from Lydia in Lesser Asia; whence its name.

When polished, it is used to try gold and silver upon, by a comparison of colour, and has thence obtained the familiar name of the *Touchstone.*

Striped Jasper by some is considered as a variety of agate or jasper, by others, of flinty slate; it sometimes shews a tendency to a slaty structure; it occurs in bands or stripes of various shades of yellow, green, purple, and red; from which it has obtained the familiar name of *Ribbon Agate* or *Ribbon Jasper.*

It is found in considerable beds. It occurs in Saxony, the Hartz, and in Sicily; and forms whole hills in Siberia.

WHETSLATE.

It is found massive, with a slaty structure, and is most commonly of a greenish grey colour, sometimes yellowish or brownish grey; it is translucent on the edges, yields to the knife, and is somewhat unctuous to the touch.

It occurs in primitive mountains at Lauenstein in Bayreuth, in Saxony, and near Freyberg in Bohemia; it was first brought to Europe from the Levant. When cut and polished, it is used for sharpening knives and other instruments; whence its name.

ALUMINE or ARGIL.

This substance obtained the name of Alumine from its forming the base of common Alum; and that of Argil, from the Latin, Argilla, Clay, on account of its being a constituent of Clays, though rarely in a greater proportion than one-third or one-fourth; nevertheless, clays are termed *argillaceous substances*, and those rocks, of which Argil forms a notable proportion, are termed *Argillaceous rocks;* one character of which is, that they give out a peculiar odour when breathed on, that may always be regarded as a mineralogical test of the presence of Argil, whence it has been termed the *Argillaceous odour*; but as it does not belong to pure Alumine, it is considered to be owing to a combination of that substance with the oxide of iron, which generally enters into the composition of argillaceous minerals.

Alumine, when pure, is perfectly white, and is destitute of taste and smell: its specific gravity is 2.0: and it is infusible, except by voltaic electricity. It has already been said, in treating of the Earths generally, that Alumine is not a simple substance, and that Sir H. Davy has ascertained it to be composed of oxygen united with a base, *Alumium*, in the proportion of 46 of the former to 54 of the latter; but though the results afford a strong presumption that Alumine is a metallic oxide, its base has not been yet obtained in such a state as to make it a fit object for investigation.

As the precise nature of its base is unknown, Alumine is still ranked among the Earths. As an Earth, it may be said that it is never found pure. It enters largely into the composition of many earthy minerals, and in small quantity in some metalliferous ores. It is an ingredient, in a large proportion, of some of the most abundant rocks, primitive, secondary and alluvial, and is found in all soils. It is the most abundant of all the Earths, except Silex.

It occurs combined with the fluoric and sulphuric acids; and with the Alkalies, Potash and Soda.

Alumine is found in the greatest purity in Corundum and its varieties.

CORUNDUM.

The varieties included under this term, viz. Corundum, Oriental Ruby, Saphire, and Emery, are the hardest substances in nature, except the Diamond, and the most ponderous of stony substances; their specific gravity varies from 3.66 to 4.08. The saphire is the heaviest. The lamellar structure is remarkably visible in the common Corundum, which readily splits into rhomboids, of which the angles are considered to be 86° 38′ and 93° 22′. All the varieties of Corundum belong exclusively to primitive countries.

E

Common Corundum, probably from its texture, has received the name of *imperfect* Corundum; and from its hardness, that of *Adamantine Spar.* It is sometimes nearly colourless, and somewhat translucent; but more often has a greyish or greenish tint, occasionally reddish, or brownish, with a metallic chatoyant lustre; it is more rarely yellow and transparent, or black and opake. The common form of its crystal is the hexahedral prism, which rarely shews a tendency to flat triedral terminations; it occurs also in obtuse, and in acute, hexahedral pyramids. It consists of about 90 per cent. of alumine, 5 of silex, and some oxide of iron: in some varieties of Corundum, the latter does not exceed one per cent.

Adamantine spar is found in India in a granite rock, imbedded, after the manner of felspar. It is often accompanied by the fibrolite, talc, garnet, zircon, and magnetic iron. It is also found in China, nearly under the same circumstances. It occurs every where from China to Bengal; in the kingdom of Ava, and on the Coast of Malabar: its gangue, in the Carnatic, is a coarse-grained white marble. It has been found in Italy in micaceous schistus, and entering into the composition of granite in North America.

The yellow is found in Bengal; the brown, with a chatoyant lustre, on the Coast of Malabar; the black in China.

In the East Indies it is used for polishing steel, and cutting and polishing gems; but the lapidaries of Europe prefer diamond-powder, on account of the greater rapidity with which it works.

The *Oriental Ruby*, or *Oriental Amethyst*, is usually of a brilliant red colour ; sometimes nearly of a olet colour; occasionally either wholly, or in partly, colourless and transparent; often chatoyant, when it is termed *Asteria*, or *Star-stone.*

The *Saphire* varies from the preceding variety principally in respect of its colours, which are blue, yellow, or yellowish green ; when blue, it is properly the saphire ; when yellow, it is by lapidaries termed the *Oriental Chrysolite*, or *Oriental Topaz;* when yellowish green, the *Oriental Emerald.*

The Oriental ruby and the saphire do not essentially differ from common corundum in respect of analysis. A variety of the latter consisted so nearly of pure alumine, that Klaproth found only 0.5 of silex and 1 of oxide of iron. The general forms of their crystals are much the same as those of common corundum, but their planes are usually more numerous. These gems are said to have been found in granite, and in sienite, in the kingdom of Pegu and the Island of Ceylon : but they are more commonly met with in alluvial deposites, and in brooks in the neighbourhood of primitive mountains. They have been thus found in the brook Expailly in France ; near Meronitz and Billin in Bohemia : they have also been found in the province of Forez in France.

The value of these gems in jewellery is well known. It is said that those of a light blue colour, may be divested of it by heating them in a charcoal crucible, without injuring their other properties, and that they are then often sold as diamonds.

Emery, though it bears very little resemblance to the preceding, is, from its hardness and analysis, considered to be a variety of Corundum. It usually occurs in masses of a blackish or bluish grey colour, having the aspect rather of a fine grained rock, than of a simple mineral.

It is found in the East Indies, enclosing whitish or reddish talc, and small portions of magnetic iron. That of Jersey resembles magnetic iron in mass, enclosing white mica. That of Smyrna is also micaceous; and encloses magnetic iron and sulphuret of iron. In the isle of Naxos, Emery is found in rounded masses at the foot of primitive mountains. It occurs in Italy and in Spain : but that of Ochsenkopf, near Swartzenberg in Saxony, seems to be the only variety which has been seen in its native place. It is disseminated in a bed of hard steatite, of a yellowish grey or apple green colour, mixed with common talc.

It is largely used for cutting and polishing by lapidaries, and by workers in glass, steel, &c.

FIBROLITE.

The Fibrolite is white, or of a dirty grey colour; it is fibrous, and harder than quartz. The fibres are rarely so large as to present any very determinate form ; but the Count de Bournon observed some in that of a right prism with rhombic bases, of which the angles are 100° and 80°. The Fibrolite is infusible; its specific gravity is about 3.2; and it is composed of 58 of alumine and 38 of silex.

It is found accompanying crystals of corundum in the Carnatic and in China.

ROTTENSTONE.

The Rottenstone is commonly considered as a variety of Tripoli, from which it essentially differs in respect of composition. It is found at Bakewell in Derbyshire. It is of a dirty reddish brown or nearly black colour, yields to the nail, and is fetid when rubbed or scraped: it is composed of 86 parts of alumine, 4 of silex, and 10 of carbon.

PINITE.

The Pinite is found generally in six-sided crystals, sometimes modified, of a brown, blackish-brown or grey colour: externally its crystals are ochreous, and usually give out an argillaceous smell when breathed on. The Pinite consists of about 74 of alumine, 29 of silex, and 7 of oxide of iron.

It was first discovered in granite, near Schneeberg in Saxony, in the mine called Pini, whence its name: it has since been found in the Puy de Dome in France, in a porphyritic felspar: it also occurs at St. Michael's Mount, in Cornwall, in granite veins; at Ben Gloe and Blair-Gowrie in porphyry.

CYANITE, or SAPPARE.

This mineral usually occurs in lamellar oblique prisms, of a bluish or pearl-grey colour, having a pearly lustre : it scratches glass when held in one direction, but yields to glass in another direction : it becomes electric by friction. Its specific gravity is about 3.5. and it consists of about 55.5 per cent. of alumine, 43 of silex, and 0.5 of oxide of iron. It is infusible.

It is usually found in primitive rocks; and occurs in Scotland, at Bohan in Banffshire; near Banchory in Aberdeenshire; and in the Mainland, the largest of the Shetland islands : it also occurs in the Tyrol; in Siberia; and near Lyons, in France, in granite; &c.

DIASPORE.

The Diaspore is a rare mineral, having for its gangue a ferruginous clay, but nothing is known of its geological situation. It is composed of a mass of slightly curvilinear laminæ of a pearly lustre, which may be readily separated from each other. By exposure to the heat of a candle, it crackles and flies off in minute fragments with a brisk decrepitation : this is supposed to be owing to the water it contains. The Diaspore scratches glass; its specific gravity is 3.43; and it consists of 80 per cent. of alumine, 0.3 of iron, and 17 of water. It is conjectured by some to be a variety of the Wavellite.

STAUROLITE.

The Staurolite is of a greyish or reddish brown colour, and occurs usually in rhomboidal or hexahedral prisms, of which the terminal edges are sometimes replaced. Haüy has noticed seven varieties in the form of its crystals, which commonly intersect each other at right angles; and he considers the primitive to be a right rhomboidal prism of 129° 30' & 50° 30'. The Staurolite is sometimes opake, sometimes translucent, with a vitreous lustre; of about the hardness of quartz, and infusible. Its specific gravity is about 3.30; and it is composed of 52.25 of alumine, 27 of silex, 18.5 of oxide of iron, and 0.25 of oxide of manganese.

When of a reddish brown colour, and in the form of four or six-sided prisms, it has sometimes, from its resembling the garnet in colour, been called the *Grenatite.*

The Staurolite belongs to primitive countries. It has been found in Brittany, near Quimper, in a micaceous clay, considered to be the debris of a primitive rock: it occurs also at St. Gothard, imbedded in micaceous schistus; and at St Jago de Compostella in a primitive rock, and is accompanied by the Cyanite.

AUTOMALITE.

The Automalite is by some considered to be a variety of the Spinelle Ruby; and as it contains a con-

siderable proportion of the oxide of zinc it has obtained the name of the *Zinciferous Spinelle*; sometimes it is called the *Gahnite*, in honour of Gahn,
its discoverer. The specific gravity of the Automalite is 4.26—4.69 : it is therefore much heavier than
the Spinelle Ruby, from which it also differs in being
nearly opake, and of a dark bluish-green colour,
as well as essentially in respect of composition : it
consists according to Vauquelin of 42 parts of alumine, 4 of silex, 28 of oxide of zinc, 5 of oxide of
iron, and 17 of sulphur. Some mineralogists have
concluded that the Automalite is the Pleonaste
loaded with sulphuret of zinc; which amounts to
an acknowledgment that it is essentially a very different substance ; and therefore it ought not to be
considered, any more than the Pleonaste, as a variety of the Spinelle Ruby.

It is found only at Fahlun in Sweden in a talcose
rock.

CHRYSOBERYLL.

This substance occurs in rounded pieces, massive,
and crystallized ; it is of a green colour, sometimes
with a yellowish or brownish tinge, and occasionally
shews an opalescing bluish white light internally.
The general form of the crystals is prismatic; the
prisms are terminated by a variable number of planes.
It readily becomes electric by friction ; is infusible,
semi-transparent, and scratches quartz. Its specific
gravity is about 3.8. That of Brazil is composed of

71.5 per cent. of alumine, 18 of silex, 6 of lime, and 1.5 of oxide of iron. It is sometimes called the *Oriental* or *Opalescent Chrysolite.*

It is chiefly procured from Brazil, where it is found accompanying topazes; and has been noticed in sand from Ceylon, together with rubies and sapphires: a few specimens have been brought from Nerbschinsk in Siberia. Its geological situation is not known.

SOMMITE.

The Sommite usually occurs in grains, or in small regular hexahedral prisms (the form of the primitive crystal), of which the lateral edges are sometimes replaced. It is of a greyish or greenish white colour, with a shining vitreous lustre, and scratches glass. The Sommite considerably resembles phosphate of lime, but may be distinguished by its superior hardness, and by its not giving a phosphorescent light when placed on a live coal. Its specific gravity is about 3.2; and it is composed of 49 of alumine, 46 of silex, 2 of lime, and 1 of oxide of iron.

It has been found only in the cavities of the lava of that part of Vesuvius called Mont Somma; where it is accompanied by mica and idocrase.

MEIONITE.

This mineral, like the preceding, has only been met with among the substances ejected by Vesuvius,

in the cavities of white granular limestone. It usually
occurs in four or eight-sided prisms, terminated by
tetrahedral pyramids. Haüy notices three varieties
in the form of its crystals, of which he considers the
primitive to be a right prism with square bases : its
colour is whitish, or greyish white, with a shining
vitreous lustre, and it is translucent or transparent.
Its specific gravity is 3.1.

PLEONASTE.

The Pleonaste is commonly considered to be a
variety of the Spinelle Ruby ; but it is less hard,
and a little heavier, and differs from it greatly in
colour and in composition. Its specific gravity is
about 3.8 ; and it is composed of 72.25 of alumine,
5.48 of silex, 14.63 of magnesia, and 4.26 of protoxide of iron. Its general colours 'also differ from
the Ruby : the Pleonaste appears nearly black ;
but when placed between the eye and the light, it
is translucent, and green or blue. Some have been
brought from Ceylon of a sky-blue, others of a yellowish colour.

The Pleonaste principally agrees with the Spinelle
Ruby, in being chiefly composed of alumine, and in
external form ; both occur in the octohedron, which
passes into the rhomboidal dodecahedron.

The geological situation of the Pleonaste differs
for the most part from that of the Spinelle Ruby.
The Pleonaste is found, accompanied by tourmalines, &c. in the rivers and alluvial country of Ceylon. It has often been found in the cavities of the

lavas of Vesuvius, and of Somma. It is also met with in the volcanic rocks of Somma, both calcareous and granitic; and in those of Laach, near Andernach, on the banks of the Rhine. Haüy seems to be of opinion that the latter, which is commonly termed the *blue Spinelle*, should be considered as a variety of the Haüyne.

LAZULITE.

This mineral is by Jameson called the *Azurite*, and is perfectly distinct from Lapis Lazuli, which is not satisfactorily ascertained ever to have been found in a crystalline form; whereas the Lazulite is often found in quadrangular, though not very perfect, crystals of a blue colour; it is rarely massive, and then in fine grains, or in masses not exceeding the size of a hazel nut: it is translucent on the edges, brittle, and nearly as hard as quartz. It is composed of 66 per cent. of alumine, 10 of silex, 2 of lime, 18 of magnesia, and 2.5 of oxide of iron.

It occurs in Vorau in Stiria, in a gangue of quartz, in a vein passing through micaceous schistus; but the most beautiful specimens are found in the bishoprick of Salzburg.

ANDALUSITE.

The Andalusite occurs in small masses of a reddish or purplish colour, having a lamellar structure,

with natural joints parallel with the sides of a rec-
ta gular prism : it scratches quartz, and sometimes
even the Spinelle Ruby. Its specific gravity is 3.16;
and it is composed of 52 of alumine, 38 of silex, 8
of potash, and 2 of oxide of iron.

This mineral is, on the one hand, considered by
some to be a variety of felspar, which is very fusible ; but the Andalusite is infusible : on the other
hand it is considered by others, probably from its
hardness, as allied to Corundum.

The Andalusite belongs to primitive countries.
It occurs in a vein of felspar traversing granite in
Forez in France : also in granite in Castelle in
Spain ; in Aberdeenshire in Scotland; and in Dart-
moor in Devonshire : it has also been found in
Douee Mountain, in the county of Wicklow, and
at Killiney in the county of Dublin, in Ireland.

BLUE FELSPAR.

This mineral is by some considered to be a variety of Felspar, from which it differs in respect of
colour and of composition ; also, it is less fusible
and somewhat harder. In most of these respects
it likewise differs from the Andalusite, of which it
has also been considered as a variety. It occurs
massive ; it colour it pale blue or sky-blue; it has
a lamellar structure, being divisible, though with
difficulty, into rectangular prisms ; and it is not
quite so hard as quartz. Its specific gravity is 3.06;
and it is composed of 71 of alumine, 14 of silex, 3

of magnesia, 3 of lime, 0.25 of potash, 0.75 of oxide of iron, and 5 of water.

Hitherto it has only been found at Krieglach in Styria, forming part of a rock, consisting likewis of quartz, and mica, or talc.

WAVELLITE.

This mineral is most commonly to be observed in small fibres, occasionally in six-sided prisms, diverging from a common center; it is said also to have occurred in small octohedrons either perfect, or having the apices replaced. The Wavellite, when fibrous, has a silky lustre; when crystallized, a vitreous lustre; it is translucent: its colour is various; white, or greyish, greenish or bluish-white: it is harder than calcareous spar; its specific gravity is about 2.4; that from Barnstaple is composed of 71.5 of alumine, 0.5 of oxide of iron, and 28 of acidulous water, which, as it corrodes glass, by the application of heat is supposed to contain a slight portion of fluoric acid: from its being chiefly composed of water and argil, it is sometimes termed the *Hydrargillite.* The Wavellite of Brazil contains about 4 per cent. of silex.

It was first discovered by Dr. Wavel in small veins and in cavities of a tender argillaceous schistus, near Barnstaple in Devonshire; it has been since found at Stenna Gwyn, in Cornwall: on a specimen in my possession from the latter place, of which the gangue is quartz, the Wavellite is accom-

panied by small crystals of Uranium of a bright yel-
low colour: it has also been found near Cork in
Ireland; and has been brought from Brazil by M.
Mawe, in the form of stalactites.

TOPAZ.

The Topaz is found only crystallized: its general
form is prismatic, and it is variously and dissimilarly
terminated : the prism is usually striated longitudinally, and modified. It is sometimes limpid, and
nearly transparent, or of various shades of yellow,
green, lilac, and red, and translucent. It mostly
becomes electric by heat, with polarity ; it exhibits a double refraction : its specific gravity is 3.5;
the white Brazilian topaz consists of 50 per cent. of
alumine, 29 of silex, and 19 of fluoric acid : those
of a yellow colour occasionally yield a small portion
of oxide of iron. The topaz of Saxony consists of
more silex and alumine, and less fluoric acid.

The pale greenish and almost transparent topaz of
Siberia becomes electric by being heated, not by
being rubbed : the Saxon Topazes, of a pale yellow
colour, become electric by friction, not by heat ;
but lose their colour by being subjected to fire ;
the deep yellow Topazes of Brazil, become electric by heat, and red by being placed in the fire.
Haüy now considers the primitive crystal of the
Topaz to be a rectangular octohedron of 82.°2′ and
122.°42′ ; but these admeasurements are not cor-
roborated by the reflecting goniometer. The crys-

tals in my possession exhibit 37 varieties of form :
many of them are from Cornwall.

The Topaz is found almost exclusively in primitive countries of the oldest formation; chiefly in
tin veins traversing granite; in which it is sometimes found imbedded.

It occurs in the tin veins of Schlackenwald in
Bohemia, and occasionally in those of Cornwall,
accompanying tin, fluate of lime, and mispickel.
The Topazes of Cornwall are mostly small, colourless, sometimes transparent, sometimes opake and
nearly milk white : they occur in the mines in St.
Agnes, and in veins in the granite of St. Michael's
mount, and are frequently accompanied by phosphate of lime and quartz. In the valley of Dan-
neberg in Saxony, the Topaz is imbedded in a rock
together with quartz, black schorl, mica, and lithomarga, forming an aggregate which has obtained
the name of the *Topaz Rock*. In the Uralian
mountains, Topazes are found in graphic granite :
in Brazil, they occur imbedded in an argillaceous
earth, resulting, as it is believed, from the decomposition of primitive rocks.

The *Pyrophysalite* is considered to be a variety
of the Topaz ; it is of a greenish white colour, and
not quite so hard as quartz ; by heat it gives out a
greenish phosphoric light. It is found at Fahlun
in Sweden in round masses, in a granite composed
of white quartz, of felspar, and silvery white
mica ; from which these masses are separated by
talc of a greenish yellow colour: it is composed
of about 54 alumine, 34 silex, and 10 fluoric acid.

PYCNITE, OR SCHORLACEOUS BERYL.

The Pycnite is only found in six-sided prisms which are deeply striated longitudinally, and are composed of minute parallel prisms, to which the longitudinal striæ may be owing. It is usually of a dull yellowish or reddish white colour, and translucent; it may be readily broken across the prism: it scratches quartz: its specific gravity is 3.5; and it is composed, according to Vauquelin, of 60 per cent. of alumine, 30 of silex, 2 of lime, 6 of fluoric acid, and 1 of water.

It is found entering into the composition of a rock, chiefly consisting of quartz and mica, at Altenberg in Saxony: it is said also to have been met with in Bavaria.

SPINELLE RUBY.

The Spinelle Ruby is usually found crystallized, either in the form of its primitive, the regular octohedron, or its variety an acute rhomboid, or having the edges of the octohedron replaced; and occasionally in macles presenting alternate re-entering angles. It is of various shades of red, violet, or yellow; more rarely black. It scratches quartz easily, but is not so hard as the oriental ruby, from which it is readily distinguished both by its colour and crystallization. It is infusible: its specific gravity is 3.7 ; and it consists of 84.47 alumine, 8.78 magnesia, 6.18 of chromic acid; to the latter its colour is supposed to be owing.

Like the greater number of the gems, the geological situation of the Spinelle Ruby is not accurately known. It sometimes occurs with saphires and oriental rubies in the sand of rivers: it is found in the sand of the rivers of Ceylon. It has been met with in India in a lamellar carbonate of lime, enclosing red mica, sulphuret of iron, and phosphated lime ; also in a substance greatly resembling adularia.

The scarlet coloured is properly termed the *Spinelle Ruby ;* the rose red, the *Balas Ruby :* the yellow, or orange red, the *Rubicelle ;* the violet coloured, the *Almandine Ruby.*

The pleonaste and automalite are usually considered to be varieties of the Spinelle Ruby, but they differ essentially in respect of composition.

SUBSULPHATE OF ALUMINE.

This mineral has been found in two places : it is said to have occurred in others.

It was first discovered in the neighbourhood of Halle in Saxony, in small masses, immediately under the soil, accompanied by foliated gypsum and selenite. These masses are snow white, or yellowish white and opake, adhere slightly to the tongue and yield to the nail ; they are very meagre

to the touch, light and infusible. By the aid of a
glass, they appear to consist of a multitude of
transparent, prismatic crystals.

This mineral has by some mineralogists been considered as an artificial production : a suspicion
arising from the proximity of the place at which it
is discovered to the college of Halle. Jameson is
of opinion that its form is a sufficient proof of its
being a natural production, and adds that there is
no laboratory nearer to the spot where it is found,
than a quarter of a league.

The Subsulphate of Alumine has been since
discovered by M. Webster in small masses, of a
snow-white colour and opake, lying upon the chalk,
and filling up a hollow in it, at Newhaven in Sus-
sex ; it has very much the same characters as that
found at Halle ; but is considered to be purer.

ALUM.

ALUM (of La Tolfa) consists of 49 per cent. of
sulphate of alumine, 7 of sulphate of potash, and 44
of water ; and is therefore improperly termed *Sulphate of Alumine.* It crystallizes artificially in the
regular octohedron.

It seems scarcely decided whether Alum is or
is not a natural production, in so singular a manner is it enveloped in the mineral substances which
contain it, or from which it is procured, or produced. From several of these it is procured merely
by their exposure to the operation of the atmos-

phere upon them; thus far, therefore, it may be esteemed a natural production; and as it is largely met with in certain volcanic countries, it has an equal claim with many other substances to be considered as a mineral.

Alum is found in, or procured from, at least three earthy substances of somewhat different external characters, which are described as alum-earth, alum-stone, and alum-slate.

Alum-Earth is of a brownish black colour, occasionally with a glimmering lustre, owing to the intermixture of mica; its fracture is earthy, somewhat inclining to slaty; and it is light, soft, and friable. It occurs, frequently, in beds of great magnitude, in alluvial land, and sometimes in the flœtz-trap formation; and is met with in Bohemia, Saxony, Austria, Naples, Hungary, and in the Vivarais in France.

Alum is procured by lixiviation from alum-earth, which seems to be considerably bituminous, and has a strong resemblance to bituminized wood; when left exposed to a moist atmosphere, it becomes warm, and at length takes fire: occasionally, it is used as fuel.

Alum-stone is greyish, or yellowish white, of various shades: it occurs in considerable masses which are translucent at the edges, somewhat hard and brittle; it adheres slightly to the tongue, and gives out an argillaceous odour when breathed on.

It is met with only at La Tolfa, in the states of the church, in Tuscany, and in upper Hungary: at La Tolfa the alum-stone is found in large strata,

and in large masses, among compact iron-shot argillaceous limestone, and is mixed with lithomarga,
fluor, and calcareous spar; and penetrated by
veins of quartz. In Hungary, pyrites, native
sulphur, and quartz, are often found dispersed
through it.

The pure Roman Alum is prepared from it.

Alum-slate is of a greyish, bluish, brownish, or
iron-black colour, and sometimes irridescent on the
surface; its structure is slaty, and it is soft, brittle,
and has a meagre feel; by exposure to weather, it
falls to pieces, and is covered by an efflorescence,
often somewhat bituminous. By one analysis, it
consists of alumine 44, sulphuric acid 25, silex 24,
potash 3, water 4. It is found in round masses, in
beds and strata, more rarely in veins, in the newer
argillaceous schistus, and also in transition mountains. It occurs in Saxony, Bohemia, Hungary,
France; also in the valley of the lead hills in
Scotland, in the mountains near Moffat, and in
the transition rocks of the south of Scotland. At
Whitby, in Yorkshire, there are very extensive
alum works; but the nature of the earth or rock
from which the Alum is procured, or its geological
situation, has not been accurately described.

Alum can only be procured from alum-slate by
burning it.

Alum is rarely found massive; that which is met
with in the island of Melo, most nearly approaches
to this character; if taken away, it shortly appears
again, whence it is supposed to be merely close
aggregations of efflorescences. The mines of Melo

are volcanic; there are also other volcanic mines, as of Latera, near Bolsena, in Italy, also at Solfaterra : those of Tolfa, above-mentioned, are likewise by some considered to be volcanic, while others assert that the beds of La Tolfa may be traced to the Apennines. Mines of Alum, in volcanic countries, are situated in whitish or reddish friable lavas.

Alum is likewise found in the waters of a few springs, as of those of Stikkenitz in Bohemia, and of some in Hungary.

CRYOLITE.

Of this rare mineral, which has only been brought from Greenland, the geological history is not known. It occurs massive, white, or greyish white, and occasionally brown, from an admixture of iron, and may be fractured into rectangular parallelopipeds : it is not so hard as fluor spar, and is translucent ; but by immersion in water, it becomes transparent. The name of Cryolite was given on account of its easy fusibility; it fuses, and becomes liquid before the blow pipe : and even in the flame of a candle. Its specific gravity is 2.94 ; and it consists of 21 of alumine, 32 of soda, and 47 of fluoric acid and water.

LIME.

Lime has never been found pure; when so prepared by the chemist, it is white, moderately hard,
of a hot acrid taste, and infusible except by voltaic
electricity.

It is not a simple or elementary body; but a
compound, consisting of oxygen united with a base
which possesses the colour and lustre of silver; but
which Sir H. Davy has not hitherto been able to
examine: he considers it to be a metal, and has
denominated it *Calcium.* Berzelius estimates Lime
to consist of about 28 per cent. of oxygen, and 72
per cent. of calcium.

Lime is obtained artificially, by heating the various species of carbonates, till the carbonic acid
is driven off; hence the lime obtained for cements
and agricultural purposes. For nice chemical purposes, it is procured in a purer state, by subjecting to a red heat, for some hours, either the white
Carrara statuary marble, or oyster shells; the
outer coat being first taken off. From the former,
lime is obtained which is mixed only with small
portions of silex, and sometimes with an atom of
iron; the lime procured from the latter, contains
only a little phosphate of lime. These impurities
are afterwards got rid of, by chemical processes.

Lime enters into the composition of a considerable number of earthy or stony minerals, but is not found in any earthy compound in the proportion of 50 per cent. except when mineralized by an acid; thus combined, it is found in so great abundance, that some geologists have estimated that it enters into the composition of the crust of the globe, in the proportion of one-eighth of the whole.

Lime, and its natural compounds, generally speaking, are of infinite importance and utility; in which respects they are inferior to no other mineral substances, but may, on the contrary, be estimated as superior to all.

Lime is found mineralized by the carbonic, phosphoric, fluoric, sulphuric, nitric, boracic, and arsenic acids; forming carbonate, phosphate, fluate, sulphate, nitrate, borate, and arseniate of Lime; these compounds, in common with all other natural combinations of earths with acids, are, by some mineralogists, termed *Earthy Salts.*

CARBONATE OF LIME.

The numerous minerals comprehended under the term Carbonate of Lime, differ greatly in exterior characters. Scarcely more can be said of them in the general, than that they all readily yield to the knife, and that their specific gravity is below 3. Carbonate of lime occurs crystallized, fibrous, foliated, granular, compact, and earthy. When crystallized, it is termed *Calcareous Spar*, from the Latin, calx, lime ; when granular or compact, *Limestone.*

Calcareous Spar is often extremely pure carbonate of lime, and is frequently very transparent, when it is strongly double refractive. Its colours are very various, and it is found crystallized in upwards of 300 varieties of form, all originating from an obtuse rhomboid of 105°.5' and 74°·55'; this rhomboid may be readily obtained by cleavage ; and the brilliant surfaces of the fragments are well adapted to the use of the reflecting goniometer. It is not often found in the form of its primitive crystal, which nevertheless has been met with in several parts of England. It is extremely subject to that species of crystallization which is known by the term macle, or, according to Haüy, hemitrope. Calcareous spar is not so hard as fluor spar : its specific gravity is 2.7 : it consists of 57 per cent. of lime, and 43 of carbonic acid. That of Iceland, which is considered to be the purest form of carbonate of lime, is transparent, and double refractive in a high degree. It is familiarly termed *Iceland spar*, or *Double refracting spar.* Some varieties of calcareous spar, especially those from Derbyshire, become phosphorescent when laid on a hot coal.

It occurs in veins in almost every kind of rock, from the oldest, to the newest alluvial strata, and accompanies, or constitutes the gangue of, a great variety of minerals ; and sometimes it appears in strata, between the beds of calcareous mountains. It is so generally distributed, that it would be im-

possible to give a list of its localities. The rarest
and most beautiful crystals are found in the northern parts of England; from which were obtained
a very considerable number of the vast variety
described by the Count de Bournon.

Stalactitic carbonated lime, the *Calc sinter* of
Jameson, is of various colours: the most common
of which is yellowish white. Stalactites are deposited from water loaded with particles of carbonated lime, in the hollow places and caverns of mountains : the water, finding its way into these caverns
through crevices in the roof, becomes exposed to
the air ; evaporation ensues, causing the particles of lime to precipitate. These stalactites are
sometimes solid, having a lamellar structure ; but
sometimes are of a fibrous texture, radiating from
the center to the circumference, as may be observed
when they are broken. The precise mode of their
formation does not seem to be understood ; some
are extremely thin tubes, which has caused the suspicion that, at least some of those which are solid,
are filled within ; some increase externally, and are
covered by minute crystals, or are terminated in
a manner somewhat resembling a mushroom.

Stalactites are sometimes of prodigious dimension,
and very numerous ; of which the grotto of Anti-
paros, in the Archipelago ; the Woodman's Cave,
in the Hartz ; the Cavern of Castleton, and other
caverns in Derbyshire, and that of Auxelle in France,
are remarkable instances.

Some caverns have been entirely filled with calcareous Stalactite, so that it is occasionally obtained

F

in large masses; in this state it is called *Oriental Alabaster*, and is much used in statuary and in the formation of vases: it was greatly employed by the ancients. That which was brought from the mountains of Thebais, situated in Egypt, between the Nile and the Red Sea, from near a village called Alabastron, was much prized: a nearly collossal figure of an Egyptian idol, formed of this alabaster, was in the museum (ci-devant Napoleon) at Paris.

Stalactites are not always very pendulous; sometimes they are of a botryoidal form. The flat or tabular portions formed on the floors of caverns, by deposition from the water dropping from the roof, are called *Stalagmites*.

Fibrous Carbonate of lime, familiarly, from its chatoyant lustre, termed *Satin Spar*, is composed of fine parallel fibres. It occurs near Alston-moor, in the north of England, in strata from one to four inches thick, accompanied by veins of pyrites, in a brown schist. It is susceptible of a fine polish and is employed in inlaying, and in the manufacture of necklaces in imitation of pearl. Fibrous carbonate of lime is also found at Bergère in France.

Schiefer Spar, the *Slate spar* of Jameson, occurs massive, having a straight, or undulated foliated structure; it is white, with a remarkably pearly lustre, and translucent.

It is considered to belong to primitive countries. It is generally found in chlorite, with sulphuret of zinc and of lead. It occurs in the Vosges, near St. Marie aux Mines; at Bermsgrün, near Schwartzenburg in Saxony; at Kongsberg in Norway, &c. It was also found in a tin vein traversing argillaceous schistus in Polgooth mine in Cornwall, accompanied by tin and chlorite; sometimes it passed into lamellar and nearly transparent carbonate of lime.

Aphrite. This mineral is only found in a friable state, and consists of white scales of a shining pearly lustre, and very soft to the touch. It is termed by Werner, Schaumerde (Earth-foam) and is by him considered to be nearly allied to Schieffer spar : it it is usually found in calcareous rocks.

It occurs at Gera in Misnia, and more abundantly at Eisleben in Thuringia, in mountains consisting of stratiform limestone.

Agaric Mineral is of a white colour, or yellowish or greyish white; and is soft, very tender, opake, and is so light as to float for a short time on water. It is considered to be nearly pure carbonate of lime. It is found in the beds and crevices of calcareous rocks in Switzerland, where it is employed to whiten houses. It is also found near Ratisbon ; likewise at Sunderland in the county of Durham.

Granular Limestone is massive, and composed of small grains which are of a lamellar texture and brilliant lustre; but as these grains intersect each other in every direction, the lustre of the mass is only glimmering. It is of various colours; the whitest and most esteemed, from its resemblance to sugar, has been termed by the French mineralogists, chaux carbonatée saccharoïde; but it has more generally, from its important uses in the arts, obtained the name of *Statuary Marble.* Granular limestone is also called *Primitive Limestone ;* the largest grained is generally esteemed to be of the oldest formation. Its colour varies from white, through various shades of yellow, green, grey, blue, red, almost to black ; and it is sometimes clouded, spotted, or veined : it is translucent at the edges, and is very brittle. It never encloses the remains of organized bodies, but frequently contains certain other substances dispersed through its masses; as quartz, garnets, mica, hornblende, talc, actinolyte, asbestus, sulphuret of zinc and of lead, magnetic iron, &c.

Granular Limestone is found in many, if not most primitive countries ; it sometimes forms entire mountains, but more often occurs in beds. It is considered to be of contemporaneous formation with gneiss, porphyry, argillaceous and micaceous schistus, with which it frequently alternates. In the Alps, and especially the Pyrenees, examples of this are of frequent occurrence. In the peak, on the south of Bagnère, in the Pyrenees, vertical beds of granular limestone alternate with beds of granite.

The most celebrated statuary marble was found in the island of Paros, thence termed *Parian marble*; the marbles of Naxos and Tenos, were also called by the same name, being both almost equally valuable for the purposes of the statuary; the quarries of those islands are said to be quite exhausted. Parian marble is white, large grained, and considerably translucent; it was called by the ancients *Lychnites.* The celebrated statues of the Venus de Medicis, of the Venus Capitolini, of the Pallas de Velletri, and many others, are of this marble.

The statuary marble called by the ancients *Marmor Pentilicus*, was taken from quarries on a mountain called Pentelicus, near Athens: it is traversed by greenish or greyish veins, which are commonly micaceous. Of this marble, the head of Alexander, the Indian Bacchus, the statue of Esculapius, the head of Hippocrates, &c. were made. The marble of *Carrara* or of *Luni*, is of a much finer grain, and closer texture, than the foregoing; and is now usually employed by statuaries: the quarries of this marble are on the eastern coast of the gulf of Genoa. Among many other celebrated statues made of this marble, are the Antinous of the Capitol; a colossal bust of Jupiter, &c.: and Dolomieu is of opinion that the Apollo Belvidere is of Carrara marble, but the marble-merchants of Rome consider it to be of an ancient Greek marble, differing from any now known. Dr. Clarke remarks, that the Carrara marble is milk white, and less crystalline than the Parian; that the Parian is whiter and less crystallized than the Naxian; and further, that while the works executed in the Parian marble retain, with all the delicate softness of wax, the mild lustre of their original polish, those which were finished in the Pentelican marble have been decomposed; and sometimes exhibit a surface as earthy and as rude as common limestone, owing to the veins of extraneous substances which intersect the Pentelic quarries. Statuary marble was anciently brought also from many other places besides those above mentioned: it is now found in Saxony, Bohemia, Norway, Sweden, England, France, &c.

Of marbles, there is an almost endless variety. Those most esteemed for ornamental purposes, as for chimney pieces, &c. are brought from Spain, and the Pyrenees, and from Italy. The environs of Verona are quarried in every direction. Many marbles consist almost entirely of shells ; it is to be regretted that no precise account is to be found of the many beautiful varieties abounding in almost every country. Those of our own country are scarcely noticed beyond the limits of the districts in which they occur, although many varieties are admirably adapted to ornamental purposes.

In Derbyshire there are two quarries of marble of a deep uniform black colour and without shells ; one of them is situated at Hadden, the other at Ashford, and both are near Bakewell ; the former belongs to the Duke of Rutland, the latter to the Duke of Devonshire ; both the marbles are largely employed for the purposes of chimney pieces and ornaments, of which the manufactory is carried on by Brown & Co. at Derby ; who have fixed up in their ware-rooms a large slab to be used as a looking-glass ; of so high a polish are those marbles susceptible.

Near the Peak in Derbyshire, a marble is quarried, which consists almost entirely of fossil shells, chiefly of entrochi ; this marble is used for chimney pieces.

At Wetton, near Ashbourne, in the same county, a beautiful variety of marble is quarried, which is of a greyish black colour, and contains a vast number of very minute shells of a whitish colour, giving to the mass very much the appearance of porphyry.

This is used for the same purposes as the black marbles above mentioned.

Near Kendall in Westmoreland some varieties of black and grey marbles are quarried; which have some resemblance to some of the Derbyshire marbles, and are employed for the same purposes.

At Babbicombe in Torbay, in Devonshire, is quarried one of the most beautiful marbles in existence; its colours vary from a light brown to a deep red, which are finely variegated. This marble is extensively manufactured into chimney pieces in the west of England; an attempt was lately made to introduce this beautiful marble in London; but the marble not being foreign, it failed of success.

In Durham, Buckinghamshire, and other counties of England, other marbles of less note are quarried.

At Kilkenny in Ireland a marble is found of a fine black, enclosing shells of a whitish colour, which, when the marble is cut and polished, exhibit segments of circles. This marble is much used for chimney pieces and ornaments.

Two or three varieties of marble commonly found in mineralogical collections, deserve a slight notice, though somewhat out of place.

The *Verd Antique* consists of carbonate of lime imbedded in green serpentine : its geological situation is not known.

For *Ruin marble,* see Index.

The *Lamachelli marble* exhibits beautiful irredescent colours, which are sometimes prismatic internally, but more commonly of various shades of red or orange, whence it has also obtained the name of *Fire marble.* It is found in veins at Bleyberg in Carinthia. Its colours are attributed to the shells of a variety of nautilus.

The *Cottam marble,* found near Bristol, which exhibits when cut and polished, the appearance of a landscape, consists of carbonate of lime mixed with a large proportion of argillaceous matter.

Common, or *Compact, Limestone* is in great degree allied to marble; it is fine grained, solid, and generally susceptible of a polish, which, as well as its colour, is duller than that of marble. Its fracture is earthy, or flat conchoidal; more rarely splintery. Its colours are various : yellowish-white, grey, brown, reddish, or bluish, of different shades. Two or more of these colours often occur in veins, zones, bands, &c.: it frequently exhibits appearances of arborizations. It is translucent on the edges, hard, and brittle. Its specific gravity is 2.6. Compact limestone usually contains small and variable proportions of silex, alumine, of the oxides of iron and manganese, and sometimes of inflammable matter.

Some varieties of compact limestone, and of marble, properly so called, or granular limestone, are not often found enclosing organic remains; these are therefore considered to be of early formation. Werner considers marble as a transition rock, compact limestone, as a floetz rock; their age is characterized by the fossils they contain : the older contain ammonites, belemnites, gryphites, &c.; the newer, abundance of such marine shells as are now found in the sea, and the remains of fish. Both these rocks are found in thick beds parallel with each other, though rarely horizontal; more frequently, nearly vertical. They are found together, forming a chain of stratified mountains, in the Pyrenees, above 10,000 feet in height. The summits of these mountains are rarely pointed, being mostly flat and of considerable breadth, with very steep sides, to a prodigious height. These appearances are remarkable on the edge of the Alpine, and in the center of the Pyrenean chain, especially near Grenoble. Compact or granular limestone, encloses beds or masses of red oxide of iron, of sulphuret of mercury, sulphuret and molybdate of lead, manganese, oxide and sulphuret of zinc, &c. These metals are also found in veins passing through these rocks, together with lamellar carbonate of lime, iron pyrites, malachite copper, &c. Garnets and steatite are occasionally, though rarely, found disseminated in these rocks. Brongniart mentions having seen flint imbedded in compact limestone, near Bakewell in Derbyshire.

The uses of compact limestone for building, paving, &c. and when deprived of its carbonic acid, for cements and the purposes of agriculture, are well known.

The houses of Paris are built of a large grained and soft calcareous stone, which is incapable of polish, and is of a dingy white, grey, or yellowish white colour. It is found in immense horizontal beds, forming the plains south of Paris. It is a very impure limestone, and furnishes, when calcined, a very bad lime. The use to which it is put, has occasioned its receiving the familiar name of *Pierre à bâtir.* Haüy describes it under that of Chaux carbonatée grossière. Its geological situation is between chalk and gypsum; it is above the chalk, from which it is separated only by a bluish plastic clay, as may be seen at Meudon. This variety is said to be almost peculiar to France.

Swinestone, or *Stinkstone,* so called from the strong fetid odour given out when scraped or rubbed, is found massive and compact, and of various shades of grey, brown, and black. By calcination it becomes white, and burns into quick lime. The offensive odour which it gives out when scraped, is considered to be owing to its including sulphuretted hydrogen : it is commonly attributed to bitumen, which does not seems to enter into the composition of Swinestone.

The harder and more compact varieties, which receive a good polish, are used in ornamental architecture.

It is said to occur forming whole mountains; it is more commonly found alternating with strata of gypsum, or of compact limestone. It occurs in Germany, France, and most other countries : in Shropshire and Northumberland in our own : the Cliffs on each side the Avon at Clifton near Bristol are in part or wholly composed of it.

Bituminous Limestone is brown or black, which colours are supposed to be owing to the bitumen it contains : its structure is sometimes lamellar ; sometimes compact, when it receives a good polish ; when rubbed or heated it gives out an unpleasant bituminous odour; by the continuation of heat, it loses both colour and odour, and burns into quick lime.

It belongs to secondary countries, and is sometimes found in coal formations, as in Galway in Ireland, where it is employed as a combustible. In Dalmatia it is so bituminous that it may be cut like soap, and is employed in the construction of houses; when finished, they set fire to the walls; the bitumen burns out, and the stone becomes white; the roof is then put on, and the house afterwards completed.

Oolite or *Roe-stone*, so denominated from the resemblance between the little round masses of which it is composed, and the roe of a fish, is always found massive, and in beds, whose geological situation is between sandstone, common limestone, and gypsum. The globular particles are sometimes composed of concentric lamellæ, and usually adhere by means of a calcareous cement. The Roe-stone is very soft when first quarried, but hardens by exposure to the air. Its colour is whitish, yellowish white, or ash grey, depending, as it is believed, on the quantity and quality of the argillaceous matter with which it is usually combined. It is a very impure carbonate of lime, and will not burn into quick lime.

The houses of Bath are for the most part built of this mineral. The Ketton-stone, and, by some, the Portland-stone, is considered to be a variety of Roe-stone. It is also found in Sweden, Switzerland, abundantly in Thuringia in Saxony, and near Alencon in France.

It is sometimes used as a marl for agricultural purposes. It was heretofore supposed actually to consist of the roes of fishes, petrified: the cause of its singular formation is not understood. Daubenton, Saussure, Spallanzani, and others, suppose it to have originated in small grains of carbonated lime, which received additional coatings by the movement of the waters which contained it.

Pea-stone, or *Pisolite,* differs from the roe-stone both in colour and structure; it is generally white, and is composed of round or spheroidal masses, from the size of a pea to that of a hazel nut, imbedded in a calcareous cement: these masses always consist of concentric lamellæ, in the midst of which is uniformly found a grain of sand. It is less abundant than the Roe-stone.

The waters of Carlsbad in Bohemia issue from beds of the Pea-stone, which is found in the waters of the brooks that supply the baths of St. Philip in Tuscany, which suffer a whirling motion in their course. It also occurs in Hungary, and at Perscheesberg in Silesia.

Madreporite is found in large, detached, roundish masses, of a greyish brown, or greyish black colour, and opake; which are composed of cylindrical, prismatic, parallel, or diverging concretions. Its name was given from its structure and appearance. It consists of 93 per cent. of carbonate of lime, together with small portions of carbonate of magnesia and of iron, carbon, and siliceous sand.

It occurs in detached masses in the valley of Rus-
bach in Saltzburg. Some naturalists have supposed
it to be a real petrifaction, which has been doubted
by others, who are of opinion that its internal structure does not warrant the conclusion.

Chalk is usually white, occasionally greyish or
yellowish white ; it has an earthy fracture, is meagre
to the touch, and adheres to the tongue ; it is soft,
light, and always occurs massive. The purest consists of carbonate of lime and water, but it more
often contains variable proportions of alumine or
silex.

It is one of the newer secondary rocks, and
wherever found is always the prevailing substance,
forming hills of three or four hundred feet in elevation, which are remarkable for the smooth regularity of their outline. Chalk is far less abundant
in nature than compact limestone. The countries
in which it is principally found, are Poland, France,
and England ; most abundantly in the latter, forming long continuous hills, in the direction nearly of
east and west, and separated by ranges of sandstone,
and low tracts of gravel and clay.

Of Chalk there are two formations, the upper
and the lower ; the latter is without flints ; the former, whatever may be its elevation, is characterized
by containing parallel and horizontal layers of flints.
Chalk likewise contains abundance of the remains
of marine organic bodies, and of amphibious and
land animals.

The uses of Chalk are numerous : when compact
it is used for building ; it furnishes lime for cements

and manure; it is employed in the polishing of metals and of glass; by mechanics, as a marking material, and as moulds to cast metals in; by chemists and starch-makers, to dry precipitates on, for which it is peculiarly qualified by the facility with which it absorbs water. It is the white of distemper painting, and, when washed and purified, forms the substance termed whiting.

Marl. Of this substance there are many varieties: some of them effervesce strongly with acids, and are employed as manures: they vary much in respect of colour, and are greyish or yellowish, bluish or reddish. Marl, in the general, is massive, but falls to pieces by exposure to the air, when it becomes plastic in water.

Calcareous marl, of an earthy texture, occurs in beds in secondary limestone, and often contains shells; it is found in most calcareous countries; occasionally in coal formations. It is sometimes found of a slaty structure and bituminous, when it is termed *Bituminous Marl Slate,* or *Marlite;* which occurs in beds with the oldest flœtz limestone, intermixed with the ores of copper: in Thuringia extensive works are employed in the smelting of the copper it contains. It is remarkable that a large number of fish, of the same species, are also contained in this substance in regular layers; the bodies of which are carbonized, or are converted into coal, and sometimes their scales are plated with copper ore; but every fish is in a contorted position, as though it had undergone a violent death by a sudden irruption or deposition of sulphureous and metallic matter: accompanying the fish, are found petrified plants, which appear to belong to the genus, fucus.

Tufa is the most impure, the most irregular, and the most porous of all the varieties of carbonate of lime. It is light, cellular, and often incrusts other substances. The various articles which, being placed in certain springs, or waters, in Derbyshire, become covered by an earthy substance, and which thereby acquire the external appearance of petrifactions, are in fact only incrusted by a kind of Tufa. It is sometimes sufficiently massive to be employed as a building stone.

It generally occurs in alluvial land, and is found both in Essex and in Derbyshire.

ARRAGONITE.

The Arragonite, so called from its having been first discovered at Arragon in Spain, is commonly found in six-sided crystals of a greyish, or greenish white, and of various shades of brown ; sometimes of a brownish red colour. The crystals, however, are not perfect prisms; down the center of each lateral plane, there generally runs a seam, which is considered to be owing to the peculiar construction of the crystal. The Count de Bournon considers this substance as a hard carbonate of lime ; it readily scratches the common carbonate and sometimes glass ; and he conceives that the six sided prism of the Arragonite is derived from the rhomboid, which

he imagines to be the common primitive crystal of
Arragonite and the carbonate of lime ; but he shews
that the six-sided prisms of these substances cannot
be derived in the same manner from the rhomboid,
because they cannot be cleaved in the same directions. Haüy, on the contrary considers the primitive
form of the crystal to be a rectangular octohedron.
The Arragonite is sometimes seen in crystals, which
appear to be elongated octohedrons, crossing each
other at right angles.

Like the common carbonate of lime, the Arragonite possesses a double refraction ; but differs
from it, in being somewhat heavier, of an imperfect
lamellar structure, and considerably harder.

These circumstances are sufficient to render the
identity of Arragonite and common carbonate of lime
doubtful ; nevertheless Vauquelin, Klaproth, Che-
nevix, &c. have not discovered any difference in
their component elements ; but Stromeyer by three
analyses discovered from 2 to 3 per cent of strontian involved in its composition.

In Arragon, in Spain, it occurs disseminated in a
ferruginous clay, accompanied by sulphate of lime :
at Leogang, in Salzburg, in an argillaceous or a
quartoze rock, accompanied by calcareous spar,
yellow copper, and arsenical pyrites. It has also
been found in the cavities of basalt near Glasgow.
The greenish varieties are brought from Marienberg
in Saxony, and Sterzing in the Tyrol. It is also
met with at Bastan and Caupenne, in the lower
Pyrenees.

The Arragonite is also found *acicular ;* either in

slender diverging, or in parallel fibres. It occurs in radiated masses, terminated by crystals, in the fissures of a compact basalt, at Vertaison, in the department of Allier, in France.

The substance termed *Flos Ferri,* because it was originally found in mines of spathose iron, is now considered to be a variety of Arragonite; though for what reason it is difficult to say, as it has never been analyzed. It is usually of a snow white, and either in small branches which are strait, or bending in various directions, having cmmonly an external silky lustre, arising probably from the crystalline terminations of the minute fibres of which it is composed; these fibres radiate from the center, presenting when the substance is broken, a brilliant silky lustre.

The finest specimens are brought from the mines of Eisen-ertz in Stiria: it occurs also at Schemnitz, at St. Marie aux Mines, and in the mines of Baygorri and Vicdessos in the Pyrenees. Small, but beautiful specimens have also been brought from Dufton in Westmoreland.

BITTERSPAR.

Bitterspar is usually found in crystals in the form of its primitive crystal, the rhomboid, which is so nearly allied to that of the carbonate of lime, that it was considered to be the same, until Dr. Wollaston discovered the difference by means of the reflecting goniometer. Its angles arc 106° 15' and

73° 45'. The colour of this mineral is yellow, with a somewhat pearly lustre; and it is harder than carbonate of lime, is semi-transparent, and very brittle. That from the Tyrol consists of 52 carbonate of lime, 45 carbonate of magnesia, and 3 of oxide of iron and of manganese.

It is commonly imbedded in chlorite, steatite, or serpentine; and is found in the mountains of the Tyrol and of Salzburg; in that of Taberg in Sweden, and on the borders of Loch Lomond in Scotland.

A variety found in compressed hexahedrons, or in small masses of a light green colour, by Dr. Thompson, at Miemo in Tuscany, thence called the *Miemite,* occurs in the cavities of alabaster. It consists of the same elements as the former variety, in about the same proportions.

A variety in the form of somewhat oblique tetrahedral prisms, was found at Gluckbrunn in the territory of Gotha. The proportions of the component elements differ from those of the preceding varieties.

BROWN SPAR.   PEARL SPAR.

The *Brown Spar* is of various shades of grey, brown; sometimes reddish brown. It occurs crystallized in varieties of the rhomboid, which, as its primitive crystal, differs somewhat from that of the carbonate of lime or the bitterspar, as was discovered by Dr. Wollaston by means of the reflecting goniometer. Its angles are 107° and 73°. The fracture of massive Brown Spar is curved-foliated; rarely

perfectly lamellar : it is translucent on the edges :
it contains a very variable proportion of iron.
Some varieties greatly resemble spathose iron.

It is commonly found in veins, accompanied by
quartz, carbonate and fluate of lime, lead, zinc,
iron, silver, &c. It occurs in the Pyrenees, Saxony,
France, Sweden, &c. At Ormes-head in Carnar-
vonshire, it occurs in veins with copper and manganese, and very abundantly in mass.

Pearl Spar is white, greyish or yellowish white,
and occurs in rhomboids usually with curvilinear
faces; sometimes of a pearly lustre which is remarkably brilliant, from which it obtained its name : it
occurs in nearly the same places and under the
same circumstances, as Brown Spar; and is abundant
in some of the mines of the north of England. That
of Sweden, consists of 29.97 of lime, 21.14 of magnesia, 44.6 of carbonic acid, 3.4 of iron, and 1.5 of
manganese.

DOLOMITE.

The Dolomite mostly occurs massive, but is sometimes of a slaty texture ; it consists of fine grains,
which are lamellar ; the mass is generally white, occasionally with a tinge of yellow or grey ; it is soft,
yields to the nail, is translucent on the edges, and
when struck, mostly emits a phosphorescent light,
which is visible in the dark. It greatly resembles
primitive limestone, but is much softer. That of the
Apennines consists of 59 carbonate of lime and 40

carbonate of magnesia; that of St. Gothard contains some oxide of iron and of manganese.

It occurs only in primitive mountains, in veins or beds, accompanied by iron, zinc, orpiment, yellow copper, mica, &c. To the intermixture of this latter substance in Dolomite, its occasional slaty texture is owing. It is found at Mont St. Gothard in the Alps; and at Simplon in the valley of Kanter; and in large veins traversing granite near Varallo, in the valley of Sesia: it also occurs in Siberia.

MAGNESIAN LIMESTONE.

The Magnesian limestone differs from common limestone in its external characters, in having generally a granular, sandy, structure, a glimmering or glistening lustre, and in being of a yellowish colour. It consists of about 30 of lime, 21 of magnesia, 47 of carbonic acid, 1 of clay and oxide of iron.

It occurs in strata at Bredon hill near Derby; at Matlock in the same county. A great range of hills extending from Nottingham to Sunderland, overlaying the coal, are entirely composed of it; it forms beds in the Mendip hills in Somersetshire; it occurs at Ballyshannon in Ireland, and at Houth, near Dublin. The Minster and city walls of York are built of magnesian limestone; sometimes, though rarely, it contains shells, &c.

The lime obtained from it is greatly esteemed for cements, being less subject to decay, owing to its absorbing less carbonic acid from the atmosphere

than the lime of common limestone. But for agricultural purposes it is less esteemed ; when laid on
particular soils it tends rather to injure than to improve vegetation ; which is wholly destroyed when
the quantity is large : this effect is owing to the
magnesia it contains. An immense tract of chalk
in France is wholly divested of vegetation, owing to
its containing about 11 per cent. of magnesia.

LIAS. CALP. ARGILLO-FERRUGINOUS LIMESTONE.

Argillo-ferruginous Limestone is found massive
in beds, or in globular and spheroidal masses, traverses by veins of calcareous spar. It is tougher
than common Limestone, and is of a bluish black,
(blue Lias) or greyish blue colour (white Lias) ;
it has an argillaceous odour when breathed on, snd
when burnt is of a buff colour. Calp is composed
of 68 per cent. of carbonate of lime, 18 of silex,
7.5 of alumine, 2 of oxide of iron, 3 of carbon
and bitumen, and 5 of water.

It is quarried at Leixlip near Dublin, (Calp of
Kirwan) and occurs in beds at Aberthaw in Gla-
morganshire, whence it has obtained the familiar
name of *Aberthaw Limestone.* The name of Lias,
which originally was provincial, has of late been
much adopted by mineralogists. The blue and
white varieties alternate with each other, generally,
in thin beds. The Lias encloses ammonites, and
great variety of sea shells ; and is remarkable for
containing the remains of crocodiles at Lyme in

Dorsetshire. Its geological situation is under the Oolite, as near Bath, and above the red marl, as in some parts of Somersetshire. It occurs in sphe-roidal masses in the blue clay of the Isle of Shep-pey, and of Highgate Hill, &c. When burnt, it forms a cement, which has the property of setting very strongly under water, and for this reason was used in constructing the Edystone Lighthouse.

Lias has of late been employed in a manner which merits notice, as being a branch of the curious and important art of multiplying copies of drawings or of penmanship. A drawing is made on prepared paper with a peculiar ink. A slab of Lias, perhaps an inch thick, is then heated, the drawing is placed upon it, and both are passed through a rolling press. The paper is afterwards wetted, and washed from off the stone; but the ink, being of a gummy or glutinous quality, becomes in part absorbed by the stone, and remains. It is then ready for the printer. Previously to the taking of each impres-sion, fresh ink is added; but the stone is first wetted with a sponge, in order to prevent the ink, which is said considerably to resemble printers ink, and to be put on with a ball similar to that used by letter-press printers, from adhering to it: the con-sequence is, that it adheres only to the ink absorbed by the stone from the paper on which the drawing was originally made: paper is then placed on the stone, and both are passed through a rolling press as before. This art has been practiced in Germany with great success, though, it is said, not precisely in the same manner; the practice there, being to make the drawing upon the stone, with a prepared ink, whence it may correctly be termed the *Lithographic Art;* the Lias of that country, is particularly adapted to it; some beautiful specimens of this art may be seen in this country. It is also said that copies of military drawings and orders were multiplied by this means, to a very large amount, at the head-quarters of the armies lately employed on the continent. An artificial composition is sometimes used instead of the Lias.

APATITE. PHOSPHATE OF LIME.

Apatite is both harder and heavier than the carbonate or fluate of lime. It occurs massive, and crystallized in the six-sided prism, (which is the form of its primitive crystal), variously terminated. The crystals in my possession, exhibit 27 varieties of form, which are extremely beautiful, and were principally brought from Cornwall. The Apatite is white, or of various shades of green, blue, red, or yellow, but not brilliant: its specific gravity is about 3. and it is composed of 53.73 of lime, and 46.25 of phosphoric acid.

The *crystallized* is chiefly met with in the veins of primitive mountains, especially in those containing tin; and it accompanies quartz, fluate of lime, sulphate of barytes, felspar, wolfram, &c. It is thus found in the mines of Saxony and Bohemia. It occurs in St. Gothard in a chlorite rock, with adularia and mica: at Stenna-Gwyn in Cornwall,

in yellowish or greenish talc : near Nantes in
France, it is met with in granite ; and in mount
St. Michael in Cornwall, in the fissures of granite,
accompanied by oxide of tin and topazes.

The Apatite is commonly phosphorescent by
heat, and it is remarkable that the prisms of such
crystals as are not phosphorescent are terminated
by six-sided pyramids, like crystals of quartz, but
are less acute ; the others are terminated by planes ;
and it is also remarkable that those which are not
phosphorescent, have only been found in volcanic
products : in those of Vesuvius, they accompany
the idocrase ; they are found at Cap de Gate in
Spain, in a cellular stone, resembling lava. These
crystals were heretofore termed *Chrysolites;* they
are of an orange brown, or asparagus green ; whence
they are sometimes called *Asparagus-stone.*

Massive phosphate of lime is of a granular,
fibrous, or earthy texture, and sometimes encloses
a small portion of carbonate of lime. The fibrous
variety is found at Schlackenwald in Bohemia, in
radiated masses in tin veins : in the same veins also
occur round masses, which are granular, sometimes
even compact, and are phosphorescent by heat.
Near Truxillo in Spain, this latter variety forms
entire hills, traversed by beds of quartz. It consists principally of lime and phosphoric acid, but
also contains small portions of fluoric acid, silex,
oxide of iron, and water.

FLUOR. FLUATE OF LIME.

Fluor is found both massive and crystallized;
the latter has a perfectly lamellar structure, and
may be cleaved with great ease into the form of the
regular octohedron, which is that of the primitive
crystal. The crystals are found passing into the
cube, the acute rhomboid, the dodecahedron with
rhomboidal planes, and the regular tetrahedron :
those, in my possession, exhibit 46 varieties of
form, which are extremely interesting. The colour
of Fluor varies from the perfectly white and transparent, through various shades of blue, green, red,
yellow, and purple, almost to black : when pounded, and thrown on a live coal, Fluor gives out a
phosphoric light ; when thrown, in mass, into the
fire, it decrepitates and flies. It is harder than calcareous spar ; its specific gravity is about 3 ; and it
is composed of 67.75 of lime, and 32.25 of fluoric
acid, according to Klaproth ;—a variety analyzed
by Scheele afforded 27 per cent. of water.

The varied colours of Fluor formerly gave rise to
the now exploded names of false sapphire, false
emerald, false amethyst, false ruby, and false topaz.

Crystallized Fluor is found at Mont Blanc and
St. Gothard ; in Saxony, Germany, and in many
other countries, it occurs in veins in primitive
mountains; and accompanies oxide of tin, mica,
apatite, and quartz, in Cornwall, and at Zinnwald
in Bohemia. It occurs in argillaceous schistas
in Cumberland and Durham, with iron ore, quartz,

calcareous spar, and sulphate of barytes : in Derbyshire in secondary limestone, with the last-
mentioned substances, together with clay and bitumen : in limestone with galena at Beeralston in
Devonshire : it also occurs in Aberdeenshire and in
Shetland.

In the Odin mine, near Castleton in Derbyshire,
Fluor is found in veins, in detached masses, from
three inches to a foot in thickness ; their structure is
divergent, and their colours, as grey, yellow, blue,
brown, are generally disposed in concentric bands :
of this variety, called *blue john* by the miner, are
made beautiful vases, obelisks, &c. by Mawe & Co.
of Derby. Fluor is no where else found adapted
to these purposes.

Compact Fluor is harder than common Fluor,
and is sometimes of a granular texture ; in general,
it is translucent only on the edges : when placed on
a live coal, it mostly gives out a green light : some
specimens in my possession from Pednandrae mine
in Cornwall, exhibit lights of various shades of
green, blue, violet, and red.

Chlorophane is esteemed to be a variety of compact fluor ; of which it has not perfectly the aspect.
It is usually of a pale violet colour, and translucent.
It does not fly in the fire, but gives out a phosphorescent light of a most beautiful emerald green colour ; a specimen in my possession, from Pednandrae, gives out this light when placed in the flame
of a candle ; but Pallas mentions a specimen from
Siberia, of a pale violet colour, which gave a white
light merely by the heat of the hand ; by the heat

of boiling water, a green light; and when placed
on a live coal, a brilliant emerald light, that might
be discerned from a long distance.

Fluate of lime is commonly found in veins; sometimes in beds, but not of considerable extent; it
never forms mountains, and is less abundant in
nature than sulphate of lime, and very much less
than carbonate of lime. The variety termed Chlorophane only has been found entering into their
composition of primitive rocks : it occurs in granite
in Siberia. Fluate of lime sometimes fills veins almost entirely.

Fluate of lime is principally used in the reduction
of metalliferous ores, as a flux ; whence its name.
The fluoric acid has been used for etching on glass,
in the same manner as nitric acid is used upon copper. From glass plates thus engraved, a considerable number of impressions have sometimes been
taken, by great care.

ANHYDROUS GYPSUM.

It is sometimes found in eight-sided prisms, but
more often massive; it is lamellar, and may be
cleaved into the form of a right rectangular prism,
which therefore is the primitive form. Its colours
are milk white, rose, violet, or bluish; it is semi-
transparent, with a double refraction, and is harder
than common gypsum ; it scratches calcareous spar.
When pure it consists of 40 per cent. of lime, and 60
of sulphuric acid ; it is sometimes called the *Anhy-*

drite, in reference to its being without water. Occasionally it yields a variable proportion of muriate
of soda, which has occasioned its obtaining also the
name of *Muriacite.* The former has only been found
in the salt mines of Bex in the Canton of Berne in
Switzerland; the latter only in those of Halle in
the Tyrol; but in fibrous, or globular masses, or
in ramose contortions *(Pierre de trippes)* it is found
in some of the mines of Saxony and of Derbyshire.

Anhydrous gypsum, affording by analysis 8 per
cent. of silex, and having the compact texture of
certain varieties of marble, is found at Vulpino in
Italy. It is of a greyish white colour, with bluish
grey veins, and is translucent on the edges. At
Milan it is employed for tables and chimney-pieces,
under the name of *Marbre Bardiglio di Bergamo.*

GYPSUM. SELENITE. SULPHATE OF LIME.

This mineral is found crystallized, fibrous, massive, and earthy. The crystallized is generally
called Selenite; the amorphous and earthy, Gypsum: but these terms are sometimes used indis-
criminately. The primitive form of its crystals, of
which Haüy has noticed 5 varieties, is a rhomboidal
prism, of 113° 8' and 66° 52', terminated by oblique-
angled parallelograms, into which the crystals may
with care be reduced by fracture; the natural joints
are very visible: the crystals are generally transparent, with a shining pearly lustre; and are of
various shades of white, yellow, grey, brown, red,

or violet colour: sulphate of lime my readily be distinguished from carbonate of lime ; it is much softer, and yields easily to the nail : its specific gravity is about 2, and it is composed of 32.7 per cent. of lime, 46.3 of sulphuric acid, and 21 of water.

Crystallized Selenite is found at Alston in Cumberland, and in great abundance at Shotover hill in Oxfordshire. Selenite is most commonly met with disseminated in argillaceous deposites ; not often in veins : but it is said to have been met with in a vein of yellow copper ore, traversing a primitive mountain, near Nusol in Hungary : in a lead vein in Bohemia ; and in the silver mine of Seinénofske in the middle of the Altaic mountains in Siberia.

It occurs in remarkably long slender fibres, which are generally associated and curved (*Plumose Gypsum*) ; it is found in Derbyshire, and in some of the mines of the Hartz and of Hungary. At Matlock in Derbyshire, Gypsum occurs also in straight fibres of great brilliancy, of which the cross fracture is lamellar, and of remarkable lustre.

When *massive*, sulphate of lime is termed *alabaster*, but is readily distinguished from that variety of carbonate of lime which has obtained the same name ; the former yields to the nail, the latter does not. It is either granular or compact ; the granular is composed of little lamellar masses, intersecting each other in every direction ; the compact has a lamellar structure, and is found in the form of stalactites, at Mont Cenis, and other places.

Granular massive gypsum is found overlaying the

most recent of the primitive rocks, and sometimes, it is said, is enclosed by them : its colour is mostly white ; and it has been found mingled with mica, felspar, and serpentine, in Siberia ; but encloses neither argillaceous matter nor organic remains : it seems therefore to have some claim for being considered as a primitive rock.   In the Levantine valley near St. Gothard, it occurs between two beds of gneiss, and also at Bellinzina in the Alps : granular gypsum also occurs near Mont Cenis, and at Moutier near Mont Blanc.

Gypsum is also found accompanying carbonate of lime, and abundantly overlaying the rock salt deposites : it covers transition rocks in Scotland.

A posterior formation of gypsum, for there appears to be at least three formations, is found in horizontal beds, and is more intermingled with marl, and frequently encloses organic remains both of plants and animals, sometimes of birds, as surrounding Paris, and in other places in France.

Earthy Gypsum has very much the appearance of chalk, but is of a looser texture.   It occurs near Zella and Œpitz in Saxony, and is employed as a manure.   It is constantly deposited by water in the crevices of gypseous mountains.

It also occurs in efflorescences, or in round fibrous masses, sometimes in stalactites, in the lavas of the Isle of Bourbon, and of Solfatara.

Gypsum is diffused through the water of almost every spring, to which it gives (in common with other earthy salts) the property of hardness, as it is usually termed.

Gypsum sometimes forms hills.   It abounds in Switzerland, Italy, the Tyrol, in Bavaria, Thuringia, Poland, Spain; and in Derbyshire, Yorkshire, and Nottinghamshire in England; and in Pennsylvania in North America.

The uses of Gypsum are very extensive: the variety called alabaster is employed by the architect for columns and other ornaments, being more easily worked than marble; it is also turned by the lathe into cups, basins, vases and other similar articles. The manufacture of these articles in gypsum is carried on by Browne & Co. of Derby, to a considerable extent.   Alabaster is found in the neighbourhood, both white, and with veins of a reddish brown colour.   The large columns employed in the building of the elegant mansion called Kedleston Hall, which is in Derbyshire, are of the variegated alabaster of that county.   When sulphate of lime or gypsum, is subjected to a certain heat, it loses what is termed its water of crystallization, and is converted into a fine powder called *plaster of paris*; the uses of which, when beaten up with water into a paste, for taking casts of gems and statues, are well known.   In some countries, especially in North America, it is largely employed as a manure.

GLAUBERITE.

This singular and rare mineral has only been found at Ocagna, in New Castille in Spain, disseminated in rock salt.   It occurs crystallized, in the form of an oblique prism, whose alternate angles are 104° 30' and 75° 30', and whose lateral planes are transversely striated, but the terminal planes, which are of a rhombic form, are smooth and brilliant; its colour is redish yellow or grey; it is transparent, and less hard than calcareous spar, but harder than gypsum : its spec. gravity is 2.7. and it is composed of 49 per cent. of sulphate of lime and 51 of sulphate of soda.   It therefore contains no water of crystallization; when immersed in water, it becomes opake.

NITRATE OF LIME.

Nitrate of Lime is rare as a natural production; being only found in silky efflorescences on old walls, in caverns, or on calcareous rocks, in the neighbourhood of decayed vegetable matter; and in some mineral waters. Its taste is bitter and disagreeable; when prepared artificially, it is obtained in six-ided prisms, terminated by six-sided pyramids.

DATHOLITE.

This rare mineral has only been found at Arendahl in Norway; of its geological situation nothing is known, but some specimens have been accompanied by greenish talc.   The Datholite is greyish white and translucent; it has been found only in ten-sided prisms, of which the primitive form is, according to Haüy, a rhomboidal prism of 109°. 28

and 70°. 32 . terminated by rhomboidal planes. The analyses of this mineral differ a little ; according to Vauquelin, it is composed of 34 of lime, 21.67 of boracic acid, 37.66 of silex, and 5.5 of water.

A variety of this substance called the *Botryolite*, which is also found at Arendahl, occurs in concentric layers composed of very slender fibres ; it consists of 39.5 of lime, 13.5 of boracic acid, 36 of silex, 6.5 of water, and 1 of oxide of iron.

PHARMACOLITE.

The Pharmacolite is found in minute fibrous, or acicular crystals, of a white, grey, yellowish, or purplish colour ; these crystals are aggregated into globular masses, or disseminated on the sides of a vien.

It is extremely rare ; having only been found at two places : at Wittichen, near Furstemberg in Germany, it is disseminated on silky or roundish masses on granite, in a vein containing cobalt, barytes, and sulphate of lime. Its purple colour is attributed to cobalt. At St Marie-aux-Mines, in the Vosges, it is found perfectly white : its specific gravity is 2.6, and it consists of 25 per cent of lime, 50.54 of arsenic acid, and 24.46 of water.

MAGNESIA.

Magnesia is a light earth of a perfect whiteness, and is absolutely insipid; it is infusible except by voltaic electricity. It consists of oxygen united with a base *Magnesium*, which is but imperfectly known, but which is considered to be a metal, and is of the same whiteness and lustre as the bases of some of the other earths. Berzelius states magnesia to consist of about 38 per cent. of oxygen, and 62 of magnesium.

Magnesia is not, like Silex and Alumine, found in very large quantity, either nearly pure, or entering, in very great proportion, into the composition of numerous and abundant earthy substances: it is found in about thirty, in different proportions; but in most of these, magnesia is not the prevailing ingredient, though in several it exceeds 25 per cent. It is involved in a few metalliferous minerals in small quantity. It occurs combined with the carbonic, sulphuric, and boracic acids; but is found in the greatest purity in the mineral which is termed native magnesia.

NATIVE MAGNESIA.

This rare mineral has been found only at Hoboken in New Jersey, in veins from a few lines to a few

inches thick, traversing serpentine in every direction.
Its colour is white, or greenish white, with a pearly
lustre.   It occurs in laminæ, which have a laminated texture, and are frequently disposed in a radiated position.   It is semitransparent, but becomes
opake by exposure; is somewhat elastic, adheres
slightly to the tongue, and is soft : its specific gravity is 2.63. and from the analysis of Dr. Bruce, it
appears to consist of magnesia 70, and of water 30
per cent.

A variety analyzed by Vauquelin, yielded 2 per
cent. of silex, and $2\frac{1}{4}$ of iron.

CHRYSOLITE.

The Chrysolite occurs in angular, or in somewhat
rounded masses, or crystallized, usually in compressed eight - sided prisms, which are variously terminated; their primitive form, according to Haüy,
is a right prism with rectangular bases.   The colour
of the chrysolite is yellow, sometimes mixed with
green, or brown; it is transparent, and possesses
double refraction ; it scratches glass : its specific
gravity is 3.4; and it consists of 50.5 per cent. of
magnesia, 38 of silex, and 9.5 of oxide of iron.

It is found near Schelkowitz in Bohemia, and at
Jurnau, in the Circle of Bunzlau ; in serpentine, at
Leatschau in Hungary: in the river St. Denis, at
the foot of the volcano of the isle of Bourbon ; and
in the debris of the volcano of Bolsano.   The chrysolite of commerce is brought from the Levant ; it

G 6

is in little masses which appear to be rounded by attrition; but nothing is known of its geological situation.

The *Olivin* is considered to be a variety of the chrysolite, though it differs in respect of analysis; the proportions of silex and alumine contained in it being nearly reversed, and it contains a trace of lime: its external chracters agree in many respects with those of the chrysolite, but it is never found crystallized; both are by some considered to be of volcanic origin; but the correctness of this opinion, in respect to both of them, may, from their occasional geological situation, be doubted. The olivin is chiefly found in little semi-transparent masses, which, sometimes, from their being in a state of decomposition, have an irridescent and somewhat metallic lustre; it is found principally in basalts and lavas.

It occurs in basalt near the village of Colombier in the Vivarais; in the basalt of Bohemia; o Kalkberg in Russia; of Hungary; and in masses of considerable size in that of Unkel on the banks of the Rhine, near Cologne. It is also found in the same kind of rock at Teesdale in Durham; in the county of Donegal in Ireland: near Arthur's Seat, Edinburgh, and in the isle of Rum. It is found in the lavas of Etna, and of Piperino near Rome.

The semi-transparent yellowish substance enclosed in the mass of native iron found in Siberia by Professor Pallas, is generally considered to be a variety of olivin. It consists according to Klaproth, of 41 per cent of silex, 38.5 of magnesia, and 18.5 of iron.

SERPENTINE.

Serpentine is always found massive; it is trans-- lucent at the edges, somewhat unctuous to the touch, and in general, yields easily to the knife. Serpentine varies exceedingly in respect of colour; which, generally speaking, is green of various shades, or bluish green, yellowish, or redish : sometimes its colour is uniform ; more often, it is spotted or veined with various colours ; when thus variegated, it is considered to be less pure and of more recent for- mation, than when of one colour : the latter consists of 37.24 of magnesia, 32 of silex, 10.2 of lime, 0.5 of alumine, 0.6 of iron, and 14 of water.

The more ancient serpentine is ranked among primitive rocks : it accompanies, is mixed with, or alternates with, primitive granular limestone, rest- ing upon gneiss, or micaceous schistus, hence it has been called *Primitive* serpentine ; by some *Noble* serpentine. It occurs in horizontal beds on the summit of Mont Rosa ; the greatest elevation at which it has been observed. The more recent for- mation, sometimes called *Common* serpentine, is considered to be a transition rock ; it often encloses steatite, talc, asbestus, chlorite, mica, garnet, mag- netic iron, &c ; but it rarely includes metalliferous veins or beds. In the large serpentine tract of Cornwall, native copper has been found dissemi- nated.

Serpentine occurs on the side of the Alps towards Genoa; as Zœblitz in Saxony; in Bohemia and Hungary; at Dobschau in Transylvania; at Zillerthal in the Tyrol; in the Milanese; in Piedmont, alternating with beds of magnetic iron; in Spain and France. In a word, serpentine, though less abundant than many other rocks, is met with in most mountain chains. Primitive serpentine of uncommon beauty is found at Portsoy in Scotland: that of Cornwall is considered to be of more recent formation.

It is fashioned by the lathe into vases for ornamental purposes at Zœblitz in Saxony, and at Bareuth; it is also made into chimney pieces, which are very beautiful.

Steatite and *Potstone* are considered by mineralogists to be nearly allied to serpentine.

CARBONATE OF MAGNESIA.   MAGNESITC.

Carbonate of Magnesia was heretofore considered as pure *Native* magnesia, until, by analysis, the presence of carbonic acid was detected: it has nevertheless been asserted of some varieties of this mineral, particularly that of Castellamonte, that when first brought from the quarry or mine, it contains no carbonic acid, which afterwards it absorbs from the atmosphere. It is usually found in large masses, which are sometimes cellular, and soft enough to yield to the nail which occasionally gives a polish by passing over it; but internally, it is sometimes harder than calcareous spar : it adheres to the tongue.

The carbonate of magnesia of Roubschitz in Moravia, is found in a serpentine rock, accompanied by meerschaum. It is opake, tender, and of a yellowish grey colour, spotted with black. It consists of equal parts of magnesia and carbonic acid.

It also occurs in serpentine, in veins or beds near Piedmont : that of Baudissero, is accompanied by hydrophane, and contains 15 of silex, and 3 of sulphate of lime : that of Castellamonte consists of magnesia, carbonic acid, silex and water. Both these varieties are employed in the porcelain manufactory at Piedmont.

SULPHATE OF MAGNESIA.

Sulphate of Magnesia occasionally occurs in the natural state, in the form of fine capillary crystals, on the surface of decomposing schistus, or of gypsum, or of the soil, and often in coal pits. It has been observed in the quicksilver mines of Idria ; on the surface of gypsum in the quarries of Piedmont, and of Mont-martre near Paris ; on the surface of the soil in many large tracts of Andalusia in Spain, after floods ; and in the foundation and lower walls of most of the houses in Madrid, it issues in efflorescences from the mortar, arising from its decomposition, and therefore to the injury of the buildings. Similar effects are occasionally to be noticed in this country.

Sulphate of Magnesia is also an ingredient of certain saline springs. It was first discovered in the mineral water of Epsom in Surry, in 1675, from which it is extracted by boiling, and constitutes that substance which in medicine is called *Epsom Salts.* It has been since discovered in the mineral water of Sedlitz and of Egra in Bohemia. At the salt-works of Portsmouth and Lymington, it is obtained from what is termed the bittern of sea-water; which is the residue, after all the muriate of soda, or common salt, has been extracted. The mode of preparing it is not generally known.

The regular form of the crystals of this salt is a four-sided prism, terminated at each end, by a two or four-sided pyramid : its primitive crystal is a four-sided rectangular prism. The crystals shew a double refraction.

Epsom salts contain 19 parts of magnesia, 33 of sulphuric acid, and 48 of water.

BORACITE.   BORATE OF MAGNESIA.

The Boracite has only been found in the mountain of Kalkberg in the dutchy of Brunswick; where it occurs in small crystals imbedded in compact sulphate of lime. These crystals are sometimes transparent, sometimes opake, and are hard enough to give sparks with the steel. Their primitive form is the cube, which is not often found perfect; the edges and angles being mostly replaced by the planes of certain modifications to which it is subject,

and which are not common to other minerals ; four
of the angles constantly present a greater number
of facets, than the other four. The crystals become
electric by the application of heat ; manifesting the
vitreous electricity on the angles which present the
greatest number of planes, and resinous electricity
on the others. The Boracite consists of about 83
parts of the boracic acid, and 17 of magnesia : when
translucent or opake, it contains a proportion of
carbonate of lime.

ZIRCON.

Zircon, when pure, is white, rough to the touch, insipid, and insoluble in water; and is about 4 times its weight. Like the other earths, it is infusible, except by the powerful action of voltaic electricity; by the assistance of which it has been ascertained that Zircon is a compound, consisting of oxygen united with a base *Zirconium;* the nature of which is unknown. The proportions in which oxygen and zirconium enter into the composition of Zircon, have not been determined.

It is very sparingly found; and then only entering into the composition of three substances, together with silex and oxide of iron; and in one instance with a small portion of oxide of titanium: it has not been detected as a component part of any rock.

Zircon has not been put to any use.

The three minerals which are principally composed of Zircon, viz. the hyacinth, the Jargoon, and the zirconite, all occur crystallized. The form of their primitive crystal is an obtuse octohedron, but their crystals commonly have twelve planes, four of which are six-sided, and are the consequence of the replacement of the lateral solid angles of the primitive crystal, causing each of its four terminal

planes to assume a rhombic form, instead of the triangular, as in the primitive crystal. The crystals
of these substances, of which I possess about 45
varieties, resemble, in a remarkable degree, those
of the oxide of tin, which also have for their primitive crystal a flat octohedron : they are doubly refractive, when translucent, and somewhat harder
than quartz, and their specific gravity exceeds 4.
These substances are infusible, but sometimes lose
their colour by exposure to heat.

HYACINTH.

The Hyacinth is of various shades of red, passing
into orange red : it is transparent or translucent:
its structure is lamellar, which is readily discovered
in one direction.

The hyacinth, as well as the two following substances, are considered to belong to primitive countries. The hyacinth, in the form of its primitive
crystal, has been found among the corundum of the
East Indies ; but is commonly found in the beds of
rivers or of brooks. It occurs in the brook Expailly,
in Auvergne in France, in a sand that is considered
to be of volcanic origin ; and is also met with in a
sand of the same description in the territory of
Vicenza, near Pisa in Italy, and in Ceylon : it is
also found at Schelkowitz in Bohemia, and in Brazil.
That of Ceylon consists of 70 per cent. of zircon,
25 of silex, and 0.5 of oxide of iron : that of Expailly consists of less zircon and more silex.

JARGOON.

The Jargoon occurs in small transparent or translucent crystals, which are considerably prismatic, and of a greyish, yellowish, brownish or reddish colour, having frequently a smoky tinge; and in rounded masses, as well as in crystals of considerable dimension, very nearly approaching their primitive form, and of a brown colour and opake: they seem to possess no regular structure. This substance is usually called the *Jargon of Ceylon:* it is found in the sand of rivers in the middle of that island, and has been met with in granite, near Cuffel, in Dumfrieshire in Scotland. Jargoon consists, according to Vauquelin, of 66 per cent. of zircon, 31 of silex, and 2 of oxide of iron.

ZIRCONITE.

The Zirconite is of a reddish brown colour, and nearly opake: it occurs in crystals imbedded in a rock, consisting of felspar and hornblende, at Frederick-Schwerin in Norway; by one analysis, it consists of 64 of zircon, 34 of silex, 0.25 of oxide of iron, and 1 of titanium.

The yellow and smoke coloured varieties of the above substances are called by the Jeweller's Jargoons, and are said to be sometimes passed off as diamonds, when deprived of their colour by heat: the red, or orange red, they call hyacinths: the commercial value of each is inferior to that of all the oriental gems: in Norway, the zirconite when cut and polished, is employed as one of the habiliments of mourning.

GLUCINE.

Glucine obtained that name from the Greek γλυκος, signifying *sweet*, on account of the sweet taste by which its salts are distinguished. When pure, glucine is a white powder, soft, and somewhat unctuous to the touch; its specific gravity is nearly 3.

Sir H. Davy has proved that glucine consists of oxygen united with a base, *Glucinum*, of which the nature is not known. It is computed that this earth is constituted of about 30 per cent. of oxygen united with 70 per cent. of glucinum.

Glucine has only been met with combined with other substances, and then only in small quantities, and in a very few minerals, viz. euclase, beryl, emerald, and gadolinite.

EUCLASE.

The Euclase is extremely rare; it has only been brought from Peru by one traveller; nothing is known of its geological situation. It has only been met with in one form, which is so complicated, that the crystals, if perfect, would have exhibited

78 planes; by cleavage, it may be reduced to a rectangular prism, which therefore is esteemed to be the form of the primitive crystal. It is of a light green colour and transparent, and readily separates into thin laminæ; but is hard enough to scratch glass, and possesses double refraction. It has not been analyzed with accuracy; by analysis of 36 grains, it yielded about 14 per cent. of glucine, 35 of silex, 18 of alumine, and 2 of oxide of iron; Vauquelin considered the greatest part of the remainder to be water of crystallization.

BERYL. AQUAMARINE

The Beryl is of various shades of yellow, green, and blue; its most common form is the hexahedral prism, which commonly is deeply striated longitudinally: it is double refractive in a slight degree, but only when held in particular directions. It occurs in crystals of various sizes; they have been met with a foot or more in length, and 4 inches in diameter, and nearly transparent. According to Vauquelin, the Beryl consists of 14 per cent. of glucine, 68 of silex, 15 of alumine, 2 of lime, and 1 of oxide of iron.

It belongs to primitive countries: it occurs in veins traversing granite, chiefly of the variety termed graphite: its gangue is quartz, or compact ferruginous clay.

It is found in the greatest abundance and purity near Nertchink in Daouria, on the confines of

China, in compact ferruginous clay. It occurs in the Altaic chain in Siberia ; and in Persia in a vein traversing a granite mountain, and is accompanied by quartz, topaz, and crystallized felspar. It has also been found in a vein passing through granite near Limoges in France ; and near Autun, in a rock chiefly consisting of felspar : in graphic granite in Pennsylvania. The Beryl is also found in Brazil, Saxony, and the isle of Elba.

It occurs in Kinloch Raimoch and Cairngorm in Aberdeenshire ; at Dundrum in the county of Dublin ; and at Lough Bray, and Cronebane in the county of Wicklow in Ireland.

It is usually considered as a variety of the emerald, but differs from it both in hardness and composition, and mostly in colour.

EMERALD.

The form in which the Emerald usually occurs, is that of a six sided prism, which also is that of its primitive crystal ; it is occasionally modified at the terminations ; sometimes each of the six lateral edges is replaced by a plane. Its colour is a pure and beautiful green ; it is somewhat harder than quartz, but not so hard as the beryl : it never occurs in very large crystals. According to Vauquelin, it consists of 13 of glucine, 64.5 of silex, 16 of alumine, 1.6 of lime, and 3.25 of oxide of chrome ; it is supposed to be coloured by the latter substance.

The emeralds known to the ancients were found in Upper Egypt, and in the mountains of Ethiopia. The finest are now found in Peru. The mine of Manta is exhausted; the present mine is situated in the valley of Tuhca, in Santa-Fé, between the mountains of New Grenada and Popayan: Emeralds occur there in veins, or in cavities, in granite. They have also been found in some secondary countries; in which they are supposed not to have been in their original situation.

The Emerald is reckoned among the gems; and when of a fine colour, and without flaws, is highly esteemed. The large emeralds spoken of by various writers, such as that in the Abbey of Richenau, of the weight of 28 pounds, and which formerly belonged to Charlemagne, are believed to be either green fluor, or prase. The most magnificent specimen of genuine emerald was presented to the church of Loretto by one of the Spanish kings; it consists of a mass of white quartz, thickly implanted with emeralds more than an inch in diameter.

CADOLINITE. See Index.

YTTRIA.

YTTRIA, in many of its properties and appearances in its pure state, bears considerable affinity to glucine; it has the same saccharine taste, but is easily distinguished from it, inasmuch as it is nearly five times heavier than water, and by some properties discoverable only by the chemist.

It has been ascertained by Sir H. Davy that oxygen enters into the composition of Yttria : but the base, *Yttrium,* with which it is combined, has not yet been seen in a separate form ; nor have the proportions in which oxygen and yttrium respectively enter into the composition of Yttria, hitherto been decided.

In the natural state, Yttria occurs as a component part of a rare mineral substance called the gadolinite, which is brought only from Sweden, and which is so called on account of its having been first analyzed by the Swedish professor Gadolin, who named the earth Yttria, because the mineral in which it was discovered, was brought from Ytterby in Sweden.

Yttria has not been found entering into the composition of any other mineral except the Yttrotantalite.

H

GADOLINITE.

The Gadolinite is of a greenish or brownish black colour, and occurs massive, and crystallized, though not very determinately, in rhomboidal prisms. It is opake, slightly translucent, and hard enough to scratch glass : it generally affects the magnetic needle, and is composed of 54.75 of Yttria, with a trace of manganese, 21.25 of silex, 5.5 of glucine, 0.5 of alumine, 17.5 of oxide of iron, and 0.5 of water.

It is found adhering to felspar and mica, in veins principally composed of the former, traversed by other veins composed of the latter substance; and is accompanied by the rare mineral called Yttro-tantalite, noticed in the description of the metal Tantalium. It has been found only at Ytterby in Sweden.

BARYTES.

Barytes, when pure, is white, has a sharp caustic taste, and as it possesses some of the characters
of the alkalies, it has by some chemists been classed
amongst them; others have denominated it an Alkaline Earth.

Barytes consists of oxygen united with a base,
Barium, with which it is united in the proportion
of about 10 per cent. of oxygen to 90 per cent. of
Barium. This base has the appearance of a dark
grey metal, which requires considerable force to
flatten it, and has a lustre inferior to that of cast
iron; but it has not been obtained in quantity sufficient to allow of the examination of its physical
or chemical qualities: some circumstances render
it probable that Barium is four or five times heavier
than water.

Barytes has never been found pure: it is combined either with the carbonic acid, forming carbonate of Barytes, or with sulphuric acid forming
sulphate of Barytes. These compounds (or, to
use the term given to minerals composed of earths
mineralized by acids, these earthy salts) may be
readily distinguished from other earthy minerals by
their superior weight; being more than four times
the weight of water.

Though Barytes is found, thus mineralized, in considerable quantity in certain countries, it is by no means plentifully distributed, since it has not hitherto been detected entering into the composition of any rock ; nor in more than one or two earthy minerals ; it is not believed to be common in soils.

Barytes is a violent and certain poison.

CARBONATE OF BARYTES. WITHERITE.

The carbonate of Barytes is of much less frequent occurrence than the sulphate. It obtained the name of Witherite, from its having been discovered by Dr. Withering, who first noticed it at Anglesark in Lancashire, in a vein, with sulphuret of lead, and some of the ores of zinc, traversing a stratified mountain, composed of beds of sandstone, slate, and coal ; the carbonate of Barytes is chiefly found in the lower part of the vein, the sulphate nearer the surface : the carbonate occurs in this vein in globular masses, having a radiated structure. It has since been found in cellular masses near Neuberg in Stiria, and at Schlangenberg in Siberia : it is also found in a lead mine near St. Asaph in Flintshire, and in many places in the north of England: at Alston in Cumberland ; Arkendale, Welthorpe, and Dufton in Durham ; Merton Fell in Westmoreland, and at Snailbach mine in Shropshire.

It is sometimes crystallized in hexahedral prisms terminated by hexahedral pyramids, and much re-

sembles crystallized quartz. The primitive crystal
is a somewhat obtuse rhomboid of 88°6′ and 91°.54′,
according to Haüy, who describes 4 varieties to
which it is subject; it is generally white and translucent, sometimes yellowish, or brownish white,
and yields easily to the knife. Its specific gravity
is 4.3 ; and it is composed of 78 of barytes and 22
of carbonic acid.

SULPHATE OF BARYTES.   HEAVY SPAR.

This mineral is found massive and crystallized :
it occurs white and transparent or opake, and of
various shades of yellow, green, red and blue.
Occasionally it resembles carbonate of lime, but
may readily be distinguished by its superior weight,
as well as by the internal appearance of its natural
joints, parallel to the sides of a right rhomboidal
prism, (the form of its primitive crystal), which may
mostly be seen, when held up to the light. The
crystals in my possession, exhibit 111 varieties of
form, which are extremely interesting ; the angles
of the primitive crystal, into which it may readily
be broken, are according to Haüy 101°. 32′.13″.
and 78°. 27′.47″, but when taken by means of the
reflecting goniometer by clear reflections on fractured surfaces, these angles afford, 101°.42′. and
78°.18′ ; which I have no hesitation in believing,
is their true value.   It is, however, remarkable, that measurements taken by that goniometer
on the natural primitive planes, however brilliant,

neither agree with the measurements above quoted, nor with those of Haüy, nor with each other.

Heavy spar is harder than carbonate of lime, but not so hard as fluate of lime ; it possesses a double refraction when held in a particular direction : its specific gravity is 4.7 ; and it is composed of 67 of barytes, and 33 of sulphuric acid.

Finely crystallized specimens are found in the mines of Hungary, Transylvania, the Hartz, Saxony, Spain, &c. ; and in our own country, in those of Durham, Westmoreland, and Cumberland : and some have lately been found, though in small quantity, in the United Mines in Cornwall. It is said to occur in stalactites in Derbyshire : it occurs in opake and compact, and sometimes in concentric lamellæ, or in fine concentric fibres, in the same county, and is there termed *Cawk.*

A variety has been found in the mines of Saxony and of Derbyshire, in small white rhomboidal prisms, laterally aggregated in columns, which have a pearly lustre and are generally translucent. It is called *columnar* heavy spar, or *stangenspath ;* it is sometimes mistaken for carbonate of lead.

Sulphate of Barytes is also met with of a *granular* texture, somewhat resembling that of statuary marble, from which it is at once distinguished by its greater weight : it is composed of 90 per cent. of sulphate of Barytes, and 10 of silex. It is found at Pegau in Stiria, with sulphuret of lead ; at Freyberg in Saxony ; at Schlangenberg in Siberia, with malachite and native copper.

Another variety which occurs near Bologna in

Italy, thence termed the *Bolognian Stone,* in translucent pieces of a smoke grey colour, gives out when rubbed a fetid smell; which by some has been attributed to the presence of bitumen.

The sulphate of Barytes is never found forming mountains, rarely in beds; but it alternates in thin beds with spathose iron at Poratsch in Hungary. Nor is it often found in large masses: but it often occurs in considerable veins, rich in metalliferous ores, in primitive, transition, and flœtz mountains. It accompanies sulphuret of antimony in the mines of Hungary, and sulphuret of mercury in those of the Palatinate; sometimes also zinc, iron, lead, and sulphuret of copper.

The uses of sulphate of Barytes are very limited; it is sometimes used in metallurgy, to facilitate the fusion of certain metalliferous gangues. It is said that the variety called Cawk is used in the smelting of copper at Birmingham.

HEPATITE.

This mineral occurs in lamellar or globular masses, of a yellowish, brownish, or blackish colour, which give out a fetid odour on being rubbed or heated; it consists of 85.2 of sulphate of Barytes, 6 of sulphate of lime: 1 of alumine, 5 of oxide of iron, and 0.5 of carbon.

It has been met with at Andrarum, and in the silver mine of Kongsberg in Norway; at Lublin in Galicia; and at Buxton in Derbyshire.

STRONTIAN.

Strontian, when pure, is white, and possesses a caustic taste : it has a strong affinity to the alkalies.

It consists of oxygen united with a base, *Strontium*, which much resembles Barium, being of a dark grey colour, and having much the appearance of a metal, but has not much lustre.

Strontian has never been found pure; but only combined with the carbonic or sulphuric acid, forming sulphate or carbonate of Strontian : and it has only been detected in one or two instances, entering into the composition of earthly substances, and then only in very small proportions ; and as it has not been found as a component part of any rock, it may be said to be a rare earth.

CARBONATE OF STRONTIAN. STRONTIANITE.

It is of a greenish or yellowish white, or of a green colour, and is somewhat harder than carbonate of Barytes ; it occurs in radiated masses, the cavities of which are sometimes lined with acicular crystals : among which, regular hexahedral prisms have sometimes been observed.

Its specific gravity is 3.67 ; it consists of 69.5 of strontian, 30 of carbonic acid, and 0.5 of water. Occasionally, it very much resembles carbonate of barytes, but it is not quite so heavy, and is somewhat harder.

It was first discovered at Strontian in Scotland, whence its name, in a vein passing through gneiss, and accompanied by galena, heavy spar, calcareous spar, and iron pyrites : it has since been found in the lead hills ; and Humboldt discovered in Peru, a variety which is white, translucent, and radiated.

SULPHATE OF STRONTIAN. CELESTINE.

This mineral is whitish or of a delicate blue ; whence it obtained the name of Celestine : it occurs in opake masses, or fibrous, or, more rarely, crystallized ; the primitive form, according to Haüy, is a right prism of 104°.28′ and 75°.12′, with rhomboidal bases ; he has noticed 8 varieties in the form of its crystals. It is not so heavy as carbonate of Barytes, its specific gravity being only somewhat above 3 ; and it is not quite so hard as fluor : it possesses double refraction. It consists, according to Vauquelin, of 54 per cent. of strontian, and 46 of sulphuric acid.

In opake spheroidal masses, it is found at Montmartre near Paris, disseminated in beds of argillaceous marl, separating beds of sulphate of lime. This variety contains 8 per cent. of carbonate of lime, and somewhat less than 1 per cent. of iron.

The fibrous variety of a blue colour, is found in a plastic clay, at Beuvron, near Toul, in the department of la Meurthe in France; at Frankstown in Pennsylvania, of a sky blue colour : and in Egypt. The spheroidal masses above mentioned, as well as specimens from Strontian in Scotland, occasionally present minute crystals.   The best crystallized specimens are said to be found at Noto and Mezzara in Sicily, in beds of sulphur, alternating with beds of sulphate of lime : it also occurs in beds of sulphur in Spain ; in sulphate of lime, in the department of la Meurthe in France ; and in beds of ferruginous marl near Bristol, sometimes finely crystallized.

ALKALINE MINERALS.

Including such as chiefly consist of an Alkali
united with an Acid.

POTASH.

It has already been remarked, in noticing the alkalies generally, that potash is not a simple body; that it consists of oxygen united with a base *(Potassium)*, which bears a strong affinity, in certain respects, to the metals; and that it much resembles quicksilver, but is lighter than water.

Potash is constituted of about 17 per cent. of oxygen, united with about 80 per cent of potassium.

It is found in the mineral kingdom, entering into combination in at least 15 earthy compounds; amongst which are felspar and mica, two principal ingredients of the oldest of the primitive rocks; it likewise occurs in 5 others combined with soda; therefore the term vegetable alkali, as applied to potash, is not correct, although it is procured in the greatest abundance from the combustion of vegetable matter of various kinds.

Potash is likewise found combined with the carbonic and nitric acids.

CARBONATE OF POTASH.

Carbonate of potash, in its various states of preparation from different kinds of vegetable matter, is also called familiarly, potash, pearlash, or salt of tartar.

The combustion of vegetable stems, leaves ashes consisting of the earthy and metallic ingredients of vegetables, and a proportion of the carbonate of potash. The latter is dissolved out by water, and being dried, is the *potash* of commerce. This when calcined, is called *pearlash*. In England and Ireland, potash is obtained from the combustion, principally, of the common fern ; 1000 parts of which, afford about 37 of ashes, and four and a quarter of potash. In the mountainous forests of Germany, and woodland tracts of Poland and Rus-sia, it is prepared in considerable quantity. The British market is chiefly supplied from North America, where the employment of making potash is subsidiary to clearing the ground for agriculture. The *tartar* that is deposited on the sides of the casks in wine countries is, when soft, formed into masses, which being dried in the sun, are after-wards piled upon a furnace with alternate strata of charcoal ; the acid and inflammable matter of the tartar is then burned off, without fusing the alkaline part, which becomes very porous and perfectly white: it is then dissolved in hot water, and being after-wards evaporated, dried, and slightly calcined, becomes fit for sale, and is the *salt of tartar* used in medicine. It consists of 48 parts of potash, 43 of carbonic acid, and 9 of water.

It has long been a question among chemists, whether the potash obtained by the combustion of vegetables is formed by this process, or whether it

previously existed in the plant. Some of the older
eminent chemists were of the former opinion, but
the latter is now gaining ground.

NITRATE OF POTASH.

Nitrate of potash, Nitre, or *Saltpetre,* is naturally
found only in an efflorescent state, in extremely
delicate fibres, and is very abundant. There are few
countries in which it is not found; it mostly occurs
on the surface of the earth; never far beneath it.
Old walls, affording animal or vegetable matter in a
state of decomposition, dry chalky plains, or sands
containing carbonated lime, are frequently covered
by it. Argillaceous earths, or pure sand, never contain it; whence it seems probable that lime is
essential to its composition, and that during its spo-
taneous formation, it absorbs at least one of its
principles from the atmosphere.

Nitre is occasionally, though rarely, found in
water; but it enters into the composition of several
plants, as of tobacco, the sunflower, hemlock, &c.
It consists of 49 parts of potash, 33 of nitric acid,
and 18 of water.

It is found on many of the plains of Spain; and
on the chalk near Evreux in France, from which it
is gathered 7 or 8 times every year; and in the deep
grottos of Mont Homburg, in Germany. In Italy
it is afforded by the calcareous soil of Molfetta.
Hungary, the Ukraine, and Podolia, furnish Europe
with abundance of nitre. In Arabia it occurs in a

valley between Mount Sinai and Suez. Persia
affords it, and it is very common in India, especially in a large plain about 60 miles from Agra in
Bengal, which is said to have been formerly well
peopled. It is found at the Cape of Good Hope.
The mountainous regions of Kentucky, which are
calcareous and full of caverns, afford it to the inhabitants of North America. In South America, the
plains bordering the sea, near Lima, are covered
with it.

But nitre is not naturally produced in sufficient
quantity for its multiplied uses. It is therefore
procured artificially. In order to this, heaps of
rubbish, of plaster and of earth, with dung and
other vegetable matter, are placed under sheds;
these are moistened with various animal fluids,
as blood, &c. and the mass is then exposed to rot
in the air. The consequence seems to be, that the
azote disengaged by the putrefaction of the animal
matter, combines with the oxygen of the atmosphere, producing nitric acid; which, by uniting
with the potash of the vegetable matter, forms nitre.
This is afterwards purified.

Nitre is employed in medicine, the arts, and in
metallurgy, for assisting the processes of oxidating
and smelting; but its principal if not its chief use
is in the manufacture of gunpowder, for which
that imported from Egypt is most esteemed, as it
contains the least calcareous matter. Gunpowder
consists of 76 parts of nitre, 9 of sulphur, and 15 of
light charcoal.

SODA.

Soda is not a simple, elementary body, but a compound, consisting of oxygen united with a base *(Sodium)*, which possesses several characters common to the metals ; it most resembles silver, but is lighter than water. Soda consists of about $22\frac{1}{2}$ of oxygen and $77\frac{1}{2}$ of sodium.

Soda is no where found in the pure state ; it enters into combination in about 12 earthy minerals already described, in proportions varying from 1 to 35 per cent. ; and is met with in 5 others combined with potash.

Soda is found both in the mineral and vegetable kingdoms ; it occurs combined with the carbonic, sulphuric, boracic, and muriatic acids ; forming carbonate, sulphate, borate, and muriate of soda.

CARBONATE OF SODA.

Carbonate of soda is both mineral and vegetable. When mineral, it is met with either dissolved in the water of certain hot springs, as those of Carlsbad in Bohemia, and Rykum in Iceland, or in certain lakes, as in Egypt and Hungary ; or in the state of a solid salt found beneath the surface of the soil:—when vegetable, it has been found to exist ready formed in the plant called Salsola Soda, and in certain sea weeds. It is procured from both these sources. Pure carbonate of soda by analysis yields 22 of soda, 15 carbonic acid, 62 water.

Carbonate of soda is found in the natural state nearly compact, but somewhat striated, between Tripoli and Fezzan in Africa; and according to Dr. D. Munro, in a stratum only one inch thick, in contact, both above and below, with muriate of soda (common salt). It is collected to the amount of hundreds of tons annually. It is called *Trona.* It rarely reaches Europe. Trona consists of 37 soda, 38 carbonic acid, $22\frac{1}{2}$ water, $2\frac{1}{2}$ sulphate of soda.

The mineral carbonate of soda called *Natron,* is procured from the lakes of Egypt and Hungary. The former are six in number, situated in a barren valley, called Bahr-bela-ma, about 33 miles westward of the Delta. The soil consists of calcareous rock, mixed with gypsum and covered with sand. These lakes contain both the muriate and carbonate of soda, and the edges of the lakes are surrounded by a band some yards in breadth, of these substances, chiefly of the latter; but the principal accumulation is at a little distance from the bank. It is taken out and exported in that impure state.

The lakes of Hungary are four in number, and lie between Dobritzin and Groswaradin; they are much neglected. The soil is a stiff blue clay covered by white calcareous sand. The lakes are from one to

two miles in circumference, and in the winter are
full of water: about April they are generally dry,
and the saline efflorescences of natron, mixed with
a little sulphate of soda, appear; which, being gathered, reappear in three or four days; this kind of
harvest continues till towards the end of October,
when winter begins, and the lakes become full of
water. The Natron, both of Egypt and Hungary,
is imported in pulverulent masses of a dirty grey
colour.

The Natron of Egypt yields by analysis about 32
of carbonate of soda, 21 of sulphate of soda, 15 of
muriate of soda, and 32 of water. That of Hungary
is composed of the same salts, but varies in respect
of their proportions.

Vegetable carbonate of soda is of two kinds,
Barilla and *Kelp.* The former is the residuum left
after the combustion of the plants, salsola soda, sali-
cornia, &c. which are cultivated by the Spaniards
on the coast of the Mediterranean. The sea water
is occasionally admitted to these plantations. When
the seed is ripe the plant is cut down, the seed
rubbed out, and the plant is burnt in a furnace.
Kelp is made from sea weeds, principally from the
leafy fuci, vesiculosus and serratus, which grow on
rocks between high and low water marks. These
plants are gathered on the shores of Britain and
other countries from May to August, and after being
dried, are burnt in pits; during combustion the
mass becomes fluid, and when cold is broken into
large pieces, for sale. It is very impure: the proportion of pure soda contained in the mass, varies
from one and a half, to 5 per cent.

SULPHATE OF SODA.

Sulphate of Soda is found in an efflorescent state, of a yellowish or greyish white colour, or in an earthy form, or more commonly dissolved in certain mineral waters : it is of a bitter saline taste, and is most commonly met with in the neighbourhood of rock-salt or brine springs.   It consists of 27 of sulphuric acid, 15 of soda, and 58 of water.

Sulphate of Soda is found in many of the lakes of Austria, Lower Hungary, Siberia, and Russia, and in Switzerland ; near Madrid in Spain it occurs in efflorescences at the bottom of a ravine, and it is said to form an ingredient of the waters of the Tagus.   It has been found in the workings of old mines near Grenoble in France, and sometimes on old walls in the same manner as nitre.   It is also found in the ashes of some vegetables, especially of sea-weeds, of the tamarind, and of some kinds of turf; and is therefore not an uncommon substance. When purified of the iron with which it is usually tinged in the native state, or when prepared artificially, it is used in medicine under the name of *Glauber's Salt.*

BORATE OF SODA.

Borate of Soda, or Borax, is chiefly, if not only, brought from Thibet, where it is procured from a lake which is entirely supplied by springs, and is fifteen days journey from Tisoolumbo the capital. The water contains both borax and common salt, and being in a very high situation, is frozen the greater part of the year. The edges and shallows of the lake are covered with a stratum of borax, which is dug up in considerable masses, and the holes thus made are gradually filled by a fresh deposition: from the deeper parts of the lake common salt is procured. The borax in its rough state is called *Tincal,* and is brought to Europe in the form of a brownish grey impure salt; or in detached crystals about an inch in length, of a pale greenish hue, and in the form of compressed hexahedral prisms.

But it is said also to be found in the island of Ceylon, in Tartary, in Transilvania, and in Lower Saxony, and abundantly in the province of Potosi in Peru.

The purification of Borax is an art confined only to a few chemists : when pure it consists of 34 parts of boracic acid, 17 of soda, and 47 of water.

MURIATE OF SODA. ROCK SALT. COMMON SALT.

Rock Salt is found in beds or masses ; sometimes crystallized in the form of the cube, which is that of its primitive crystal, and into which pure Rock Salt may readily be cleaved ; when impure, as when its brown colour is derived from an intermixture of clay, its structure is less determinately lamellar ; its lustre is shining or vitreous ; it is either translucent or transparent, and its colour is very various, as white, grey, reddish brown, brick red, vielet, blue, and green. It yields easily to the knife; its specific gravity is 2.54 ; pure muriate of soda, ac-

cording to Berzelius, is composed of 46.55 of
muriatic acid, and 53.44 of soda. Rock Salt, according to Kirwan, is composed of 33 of muriatic
acid, 50 of soda, and 17 of water: when of any of
the forementioned colours, it is always somewhat
impure. It is sometimes, though not often, of a
fibrous texture.

Muriate of Soda is one of the most abundant substances in nature; not only is it found in large beds
and masses, but also in the waters of certain springs
and lakes; and in those of every sea. It forms
about one-thirtieth part of the waters of the Ocean.

Salt, or *Brine, Springs* are not found in primitive countries; but generally arise from the newer
secondary: nevertheless they are said never to be
far distant from the chains of primitive mountains;
they are found at the foot of the Alps, the Vosges,
the Pyrenees, the Carpathian mountains, &c.; those
of Droitwich, in our own country, are surrounded
by a brownish red sandstone, which is considered
to be the old red sandstone of Werner. So immense is the quantity which rises from the four pits,
which are sunk through various kind of soil, rock,
and gypsum, that, although that which is used bears
but a small proportion to that which runs to waste,
the quantity of salt annually made from them amount
to about 16,000 tons. Salt springs are said to arise
in countries in which no deposition of Rock salt
has been discovered; the brine commonly contains
some portions of certain other salts, as sulphate of
soda, and sulphate of lime, &c. and it has been
remarked in regard to some springs, that the quantity of brine always increases after heavy rains.

Rock Salt is commonly disposed in thick beds; either superficial, as in Africa, or of a very great depth, as in Poland : sometimes they are very high above the level of the sea, as in the Cordilleras of America, and also in Savoy ; where they are found at an elevation equal to that of perpetual snow. In Spain, Rock salt occurs in vast masses, which seem to be isolated.

Sulphate of lime, or gypsum, almost always accompanies Rock salt ; and is sometimes so impregnated by it, as to be worked as a Salt mine, as at Arbonne in Savoy.

Red or greyish clay frequently alternates in beds with Rock salt ; but blocks or masses of clay are said more often to be enclosed in it : when in beds, the clay is accompanied by sand, grit-stone, and rounded pebbles, and by compact carbonate of lime, which is sometimes fetid, sometimes bituminous. Rock salt commonly rests upon sulphate of lime, and is covered by carbonate of lime. In the beds of various substances which accompany it, are sometimes found the remains of organized bodies, the bones of elephants, and other mammalia, carbonized wood, fossil-shells, and bitumen ; and frequently masses of sulphur are found in the sulphate or carbonate of lime.

Several countries in Europe abound in Rock salt and Salt springs ; Spain, Germany, Italy, part of Russia, &c. ; in England in Worcestershire and Cheshire ; in France there are many Salt springs, but no known deposite of salt. Sweden and Norway are without salt. It is abundantly found in many countries of Asia, Africa, and America.

AMMONIA.

Ammonia, or *Volatile Alkali*, when pure, subsists in a gaseous form. It is commonly believed to consist of hydrogen and nitrogen; but some experiments of Sir H. Davy have induced the suspicion that it contains 7 or 8 per cent. of oxygen.

It is only found combined with the sulphuric and muriatic acids; forming sulphate and muriate of ammonia.

SULPHATE OF AMMONIA.

Sulphate of Ammonia has an acrid, bitter taste. It occurs in the form of stalactites, of a yellowish colour, and covered by a whitish, farina-like dust which are found in the fissures of the earth surrounding certain small lakes near Sienna in Tuscany. It consists of 40 per cent. of ammonia, 42 of sulphuric acid, and 18 of water.

MURIATE OF AMMONIA. SAL AMMONIAC.

The Muriate of Ammonia is not abundant; it chiefly appears in efflorescence of a greyish white, yellowish, apple-green, or brownish black colour; but sometimes occurs in small crystals which are not very determinate. It consists of 40 per cent. of ammonia, 52 of muriatic acid, and 8 of water.

This salt is characterized by its insolubility in water, and by the ammoniacal odour which it gives out when triturated with lime, better than by any of its external characters. It sometimes exists in the substances which enclose it, in such a manner, as to be without the reach of the eye or the feel.

It is principally found in the neighbourhood of volcanoes, sublimed in the cracks of lava, among other volatile matters near their craters. It thus occurs in the lavas of Etna and Vesuvius. At Solfatara it escapes in bubbles, which are caught and condensed in long earthen pipes. It is said to appear in efflorescences on certain rocks in Turquestan in Persia, in Calmuc Tartary, Bucharia and Siberia; in some lakes of Tuscany, and certain springs of Germany; and in some English coal, especially that of Newcastle.

Volatile alkali is obtained from Muriate of Ammonia. The Sal Ammoniac of commerce is chiefly brought from Egypt, where the soot proceeding from the combustion of the excrements of certain ruminating animals who feed on saline plants, is collected in chimnies, whence the soot is taken and placed in large glass vessels: it is then heated sufficiently to drive off the muriate of ammonia, which is sublimed, and attaches itself to the upper edges of the vessel in cakes, which are always, in

I

some degree, tinged by the soot; in this state it is sometimes preferred for use in the arts. The soot affords nearly one third its weight of Sal ammoniac.

It is obtained in various parts of Europe by the distillation of animal matter.

It is used in medicine, and in the arts.

NATIVE METALS

AND

METALLIFEROUS MINERALS:

including such Metals as occur nearly pure, or
combined with oxygen, or sulphur, or one or
more of the acids ; together with those compound
substances commonly denominated Metalliferous
ores.

IRON.

Pure Iron is of a bluish grey colour, and has a granular texture; it is hard, ductile, and malleable, and is the most tenacious of metals next to gold. Like cobalt and nickel, it is magnetic; and so readily is polarity acquired by iron, that a bar remaining a long time in a vertical position, or even approaching to it, becomes magnetic. The northern pole is always at the lower extremity. The specific gravity of iron is about 7.

Iron is an ingredient of mica, which enters largely into the composition of some of the oldest and most abundant primitive rocks; and being found in all soils, and in almost every rock, it is therefore considered to be the most generally diffused substance in nature. It has been met with in the nearly pure metallic state in considerable masses, reputed to have fallen from the atmosphere; but these masses are generally alloyed by nickel. Native Iron is also said to have been found disseminated in certain metalliferous veins. The ores of iron are very numerous; it occurs combined with sulphur, oxygen, the oxides of titanium, manganese, and chrome; the phosphoric, sulphuric, carbonic, and arsenic acids; with silex, alumine, lime, and with water.

Mineralogists are not well agreed either in respect of the names or the arrangement of the ores of Iron.

The ores of Iron, which are of a dark brown or black colour, and in which the iron is considered to be combined with a small proportion of oxygen, such as the magnetic and brown iron ores, belong chiefly, though not exclusively, to primitive countries : they often form an integral part of primitive rocks.

The Red Iron ores chiefly belong to secondary and alluvial countries; they are occasionally met with in the veins of primitive mountains, but are not found entering into the composition of primitive rocks.

The red and argillaceous varieties, but particularly the latter, it is remarkable, are generally found in the neighbourhood of coal, so essential to their reduction to the metallic state ; either resting on the coal or filling up the fissures in it. Iron ore is thus found in the collieries of Glamorgan, of Monmouth, Staffordshire, Shropshire, and of those of Carron in Scotland.

It would be vain to attempt the enumeration of the uses to which iron is put by man. Steel is an artificial combination of iron with carbon. The brown colour used in porcelain painting is oxide of iron.

An ore, in which iron is combined with alumine, is used in the making of what are termed *read lead* pencils. Plumbago, or *black lead*, is a natural compound of iron, with a large proportion of carbon.

NATIVE IRON.

Native Iron is said to have been found in veins.

Scrieber mentions having met with it in the form of a ramose stalactite covered by brown fibrous oxide of iron, mingled with quartz and clay, in a vein traversing the mountain of gneiss, called Oulle, near Grenoble in France.

Bergman cites instances of malleable iron having been found in a gangue of brown garnets, near Steinbach in Saxony.

Lehman says that it was found in a vein at Eibestock in Saxony.

Karsten describes a brown oxide of iron mingled with spathose iron and sulphate of barytes, which contained native iron disseminated through the mass. It was found at Kamsdorf; and consisted of about $92\frac{1}{2}$ parts of iron, 6 of lead, and $1\frac{1}{2}$ of copper.

NATIVE METEORIC IRON.

Native Meteoric Iron is somewhat paler and lighter than common metallic iron, and is often more malleable : it is supposed that its colour and superior malleability may be owing to a small portion of nickel combined with it, and that its lightness is owing to the very numerous minute cells observable in it. It is magnetic, flexible, and cellular; the cells are occasionally filled by a yellowish and translucent substance, of a somewhat resinous appearance, by some considered as a variety of olivin.

Meteoric Iron has been found in different quarters of the globe; in Bohemia, in Senegal, in South America, and in Siberia; of the latter we have the best account. It was found by professor Pallas on the top of a mountain, on which there was a considerable bed of magnetic iron-stone, on the banks of the river Jenisei. It weighed 1680 Russian pounds, and possessed some of the important characters of pure iron, as malleability and flexibility, and was reported by the inhabitants to have fallen from the sky. The mass found in the Vice-royalty of Peru in South America, was described by Don Rubin de Celis: it weighed about fifteen tons; it was compact externally, and was marked with impressions as if of hands and feet, but much larger, and of the claws of birds; internally it presented many cavities: it was nearly imbedded in white clay, and the country round it was quite flat and destitute of water. Meteoric Iron is alloyed by about 3 parts in the 100 of nickel; which, it is worthy of remark, is also found by analysis to be a constituent part of all those stones, which in various parts of the European Continent, in England, and in America, have been known to fall from the atmosphere, and are therefore termed *meteoric stones.*

In the imperial cabinet at Vienna there is a very considerable mass of meteoric iron, which fell from the atmosphere in 1751 at Hraschina, near Agram in Croatia, appearing in the air like a globe of fire. It consists of $96\frac{1}{2}$ per cent. of iron and $3\frac{1}{2}$ of nickel.

The Abbé Haüy is of opinion that some appear-

ances of crystallization approaching the regular octohedron may be traced in native meteoric iron.

IRON PYRITES, SULPHURET OF IRON.

Iron Pyrites is tin-white steel-grey, or of various shades of yellow, and is found in mass, and
of various shapes, as stalactitic, nodular, &c. and
crystallized. It includes several varieties; some of
which greatly resemble yellow copper ore, but may
be readily distinguished, as the latter yields to the
knife, which iron pyrites does not.

One variety, which affects the magnetic needle,
is therefore called *Magnetic Pyrites*. It is not
found crystallized; it is generàlly of a bronze yellow, but is sometimes brown. Its magnetic property is supposed to be derived from its containing
a larger proportion of iron than common pyrites;
it consists of 63.5 metallic iron, and 36.5 of sulphur; and its specific gravity is 4.5.

It is believed to belong almost exclusively to primitive countries. It is met with in Saxony, Bohemia, Silesia; at Moel Elion in Caernarvonshire, it
occurs with common pyrites in a kind of serpentine,
and near Nantes in France in limestone.

When nearly of a tin-white, yellowish, or steel-
grey, it is termed *White Pyrites*. The principal
difference between it and *Common Pyrites* seems
to be, that the former decomposes much easier. The
specific gravity of the former is 4.7; of the latter
4.8. Both are found crystallized in the cube, which

is considered to be the form of the primitive crystal: the crystals of common pyrites, in my possession, afford 38 varieties of form, which are very
beautiful and interesting.   White and common pyrites differ very little in the results of their analyses.
The latter yields about 47½ parts of iron, and 51½
of sulphur.   The former contains a rather larger
proportion of sulphur.

A variety is occasionally found containing both
sulphur and arsenic; it is of a paler colour than
common pyrites.   The arsenical ore of iron called
Mispickel is noticed under the head Arsenic, to
which it properly belongs.

Iron Pyrites occurs in almost every species of
rock.   It abounds in granite, and particularly in
argillaceous schistus.   It is never wrought as an
ore of iron, but is largely employed in the manufacture of green vitriol; and sulphur is often procured
from it by sublimation.

Magnetic Iron Ore is generally of an iron-black
colour, with a slight metallic lustre.   It is found
massive, and crystallized in some varieties of the
octohedron, which is considered to be its primitive
crystal.   It is sufficiently magnetic to take up iron
filings, and possesses polarity.   It is nearly a pure
oxide of iron.   Its specific gravity is about 4.5.

It is most commonly found in primitive countries,
generally in beds and large masses; and is accom-

panied by hornblende, granular limestone, and garnet; and occasionally by blende, all the varieties
of pyrites, fluate of lime, oxide of tin, and sulphuret of lead, &c.

The mountains called Taberg in Swedish Lapland,
and Pumachanche in Chili, are said to consist almost entirely of Magnetic Iron Ore. It is plentifully found also in Corsica, Saxony, Bohemia, Silesia,
Russia, and the East Indies. In Great Britain, in
Cornwall, Devon, and the Isle of Unst in the Hebrides.

It exists in great abundance and purity at Ros-
lagia in Sweden, where it is manufactured into the
best bar iron, so much sought after by our manufac-
turers of steel, though it affords only middling cast
iron.

Some varieties of Magnetic Iron ore, either compact, sandy, or earthy, have been found, containing
from 12 to 22 per cent. of oxide of titanium. The
sandy variety is found in angular or rounded masses,
and in octohedral crytals. It occurs at Hunstan-
ton, in Norfolk; in Agyleshire; and at Arklow,
near Wicklow in Ireland, with native gold.

RED IRON ORE.   IRON GLANCE.

Red Iron ore presents itself under many varieties
of form and colour; most if not all of which are
very feebly magnetic: and though fragile, many
of them are hard enough to scratch glass: they
all afford a red powder. Iron Glance is considered,

ɪ 6

by many mineralogists, to be an oxide of iron, but not one variety has yet been analyzed. It occurs crystallised; in lamellæ; micaceous; in scales; in globular masses, and in stalactites; also compact and ochrey.

Crystallized Iron Glance occurs under very different circumstances. It is abundantly found in the old and famous mines of the Island of Elba, in very brilliant and frequently, very large crystals; and in Saxony, Bohemia, Silesia, Switzerland, France, Norway, Sweden, and in Cornwall. It is also met with, principally in irregular and compressed crystals, in the fissures of the lavas of the Puy de Dome in France, of Vesuvius, also in the Lipari islands, and the volcano of Stromboli.

It affords an excellent malleable iron, but somewhat hard; and also a good, but not the very best, cast iron. Its specific gravity is about 5.

In *lamellæ* of a shining metallic lustre, it is met with in Caernarvonshire, and at Eskdale in Cumberland.

The *micaceous* variety is found principally with other ores of iron, and sometimes with coal; it occurs in minute scales of a redish black colour, and unctuous to the touch. It is found in the Hartz, at Schemnitz in Saxony, &c. and at Tavistock in Devonshire, and Dunkeld in Perthshire; the *scaly* variety seems to differ very little from it.

Red Hæmatites Iron ore, is of a bluish grey colour, with a metallic lustre, and passes into brownish red; it occurs in globular and stalactitie masses, which internally have a fibrous diverging structure; and are sometimes in concentric layers. Its specific gravity is nearly 5. It is found in Saxony, France, Silesia, and in England; very abundantly at Ulverstone in Lancashire; and in the forest of Dean in Gloucestershire; but Norway, and other northern countries, seem almost destitute of this variety. It affords excellent iron, both cast and malleable; most of the plate iron and iron wire of England are made of it. When ground to fine powder, it is largely employed in the polishing of metal.

The *Compact* variety is of a dark steel grey, brownish, or blood red colour, and is found in mass, disseminated, &c. and sometimes in crystals that have taken the forms and places of those of other substances. It is found in France, Germany, Norway, Siberia, and at Ulverstone in England, and occurs both in primitive and secondary countries. It affords good cast iron, and malleable though somewhat soft, bar iron.

Reddle, Red Ochre, or *Red Chalk,* is of a blood red, passing into brownish-red. It is dull, earthy, and is nearly without lustre. It generally accompanies the two preceding varieties. It is used in the arts. Near Platte in Bohemia it is smelted, and affords very excellent malleable iron.

BROWN IRON ORE.

Brown iron ore, like the preceding species, is found under several varieties of form; but does not

resemble it in being slightly magnetic: all its
varieties afford a brown powder. It is not certain
wherein it differs from the Red iron ore; most of
the varieties of the Brown Iron ore have been
proved to contain a small proportion of oxide of
manganese, and this is supposed to constitute the
difference between the two species. It also occurs
scaly; in globular masses, and stalactites; also compact and ochrey.

The *scaly* variety seems to differ in its internal
characters from the scaly variety of the red, only in
being of a colour between steel grey and clove
brown.

The *Brown hæmatites* is found under nearly the
same circumstances, but not altogether in the same
countries as the red; it yields a better cast iron
than the brown; but the bar iron of the latter is
both very malleable and very hard, probably from
the manganese which it contains; hence it yields
excellent steel. Though it is found in England, it
does not occur in sufficient quantity to be wrought.

The *Compact* variety is found accompanying the
preceding in the form of stalactites, and in masses
of various shapes. It is remarkable in respect of
its forming the substance of several petrifactions,
as of madreporites and coralites. It is dull internally.

Brown Iron Ochre also accompanies brown hæmatites. It is destitute of lustre, and has an earthy
fracture; it is tender, soils the fingers, and is of a
yellowish brown colour. It does not contain any
manganese.

BLACK IRON ORE.

Black Iron Ore is a rare mineral of a bluish black colour; and is found globular, massive, &c. It is of a fine fibrous diverging structure, and has a glimmering and somewhat metallic lustre.

The *Compact* variety seems to have been found only in Saxony, Hesse, and some other parts of Germany, in primitive and secondary mountains, accompanying the preceding species. It was long confounded with compact grey manganese; but it yields a good iron which corrodes the sides of the furnace.

Black Hæmatites Iron ore has been found only at Schmalkalden in Hesse. In form it greatly resembles the Red Hæmatites, but differs in being more of a steel grey, and of a more delicately fibrous structure.

Neither of these varieties has been analyzed.

CLAY IRON-STONE.   ARGILLACEOUS IRON-STONE.

Argillaceous Iron-Stone is of an ash grey colour, inclining to yellowish and bluish; also brown or redish brown, which last colour is usually the effect of exposure to the atmosphere; some varieties have a slaty structure; some are glimmering, others dull. It occurs in globular masses, solid or pulverulent within; and in masses of various shapes, as columnar and lenticular, and in little rounded portions about the size of peas.

The *globular* masses, consisting of concentric layers either hollow, or enclosing a yellowish brown pulverulent substance, are termed *Ætites*: externally, these masses are very compact, hard and brown : they are found in argillaceous beds in some secondary countries, and sometimes abundantly disseminated in alluvial hills ; and are occasionally accompanied by bituminized wood. Ætites consists of about 77 parts of peroxide of iron, 14 of water, 2 of oxide of manganese, 5 of silex, and 1 of alumine : it is found in Bohemia and Saxony; in France, in beds of sand and of ochre. It is frequently met with in the newer secondary rocks, and in the shale of coal formations : it occurs at Coalbrookdale in Shropshire, and Merthyr Tydvil in Wales, and in Scotland; at which places it yields an iron of a fine quality. It also occurs in Staffordshire, Yorkshire, &c.

The *Columnar* variety is of a brownish red colour, and is dull, soft, brittle, and magnetic ; it is met with in round masses and in columns. It occurs occasionally in beds of shale, above coal, and in many cases seems a pseudo-volcanic product, being accompanied by burnt clay, &c. and is met with in Germany in several places, and in the Isle of Arran. It is not a common mineral, and is never wrought as an ore of iron.

Pisiform or *Pea Iron Ore,* is mostly found in round grains of a dark brown colour, having an external polish, and internal earthy fracture, imbedded in a ferruginous, argillaceous, or calcareous cement. It is found in beds of clay, and in flat beds beneath the surface : but mostly in cavities in secondary limestone. It includes less iron, and more alumine and silex, than Ætites. It supplies very considerable iron works at Arau, near Berne in Switzerland; and the greater part of the French iron is said to be produced from this ore: but the iron yielded by it is also said to be often of a bad quality, and very brittle.

BOG IRON ORE.

Bog Iron Ore is of various shades of black and brown, and is generally of a soft earthy texture. It has obtained its name from its being chiefly met with in marshy places, or in those which have been so; it consists of about 66 parts peroxide of iron, 2 of oxide of manganese, 8 of phosphoric acid, and 23 of water. A variety which is nearly black is termed *pitchy*, and occurs massive, with a shining lustre; it contains about the same proportion of iron, combined with 8 of sulphuric acid and 25 of water. The earthy variety, which occurs in yellowish-brown cellular masses, has not been analyzed. It is found in Saxony, Prussia, Poland, and many parts of the north of Europe; in the Highlands of Scotland and the Orkneys. The acids contained in this species are supposed to have arisen from decayed vegetable matter; and it is believed that owing to their presence, the iron obtained from Bog Iron Ore is what is termed cold-short, and therefore can scarcely ever be used for plate iron, never for wire.

BLUE IRON ORE, OR PHOSPHATED IRON,

is of a pale or dirty blue colour: it occurs in slender rhomboidal prisms, which are of an earthy texture: when fresh dug it is white, but by exposure becomes blue; and has been found consisting of about 47 parts of oxide of iron, 32 of phosphoric acid, and 20 of water. It is usually found in small portions or nests in certain clays, bog iron ore, or peat. It occurs in France, Saxony, Poland, and Scotland; in Siberia in fossil shells; in the lavas of Etna; and in England, in the river mud at Toxteth near Liverpool, and in the Isle of Dogs on the borders of the Thames.

SPATHOSE IRON.

Spathose Iron is of a white or yellowish grey colour, passing by decomposition into yellow, brown, and brownish black; when translucent, it has not the appearance of a metalliferous substance. It mostly occurs in rhombs, which are rarely perfect, and is found in veins, principally, of primitive mountains; it is sometimes accompanied by sulphuret of iron, yellow copper ore, grey copper, quartz, carbonated lime, &c. It is met with in Saxony, Hungary, &c. and in many places, in sufficient quantity for working as an ore of iron. At Schmalkalden in Hesse, there is a bed of the black variety 150 feet thick, which has been worked for several centuries; and at So-

mororstro in Spain, there is a hill entirely composed
of it. It occurs in small quantities in several of
the mines of Cornwall. Spathose iron ore consists
of 58 parts of oxide of iron, 35 of carbonic acid,
4.25 of oxide of manganese, 0.5 of lime, and 0.75
of magnesia. The iron obtained from it is particularly valuable, as it may be converted into excellent
steel immediately from the state of cast iron ; the
bar iron formed from it is both hard and tough.

SULPHATE OF IRON. GREEN VITRIOL.

Sulphate of Iron is found in fibres of a whitish,
greenish, or yellowish colour, occasionally of an
emerald green colour; it occurs massive, in stalactites, and pulverulent. It is also found crystallized
in an acute rhomboid, according to Haüy of 81°.33′.
and 98°.37′, which is the form of its primitive crystal ; it presents a few varieties of form. It is met
with in most mines of sulphuret of iron, of copper,
and of zinc ; the two latter being rarely exempt
from iron.

CHROMATE OF IRON.

Chromate of Iron has hitherto been found only
massive : its aspect is somewhat metallic, and it
is hard enough to cut glass ; it is of a blackish
brown colour, and when reduced to powder, of
an ash grey. It has been found in the Uralian

mountains in Siberia; and since, in nodules and veins in serpentine, near Gassin, in the department of the Var, in France. The latter consists of oxide of iron 34 parts, chromic acid 43; alumine 20; silex 2. That of Siberia differs principally in containing 10 per cent. more of the chromic acid and 10 per cent. less of alumine: it also contains small portions of lime and of manganese.

ARSENIATE OF IRON.

This beautiful substance is rarely found massive, but mostly in cubes of various shades of green, yellow, and brown, sometimes nearly red; they are translucent, and occasionally almost transparent, but possess more of the appearance of an earthy, than of a metalliferous substance; by decomposition they become of a dull brown, lose their translucency, and at length assume the pulverulent form. The cube is considered to be the form of the primitive crystal; I possess seven varieties; and it is remarkable, in some of them, that only 4 of the 8 solid angles are replaced by planes; but in the more complex variety, each angle is replaced. This mineral is said hitherto to have been found only in some of the copper veins of Cornwall. In that called Huel Gorland, which is partly in argillaceous schistus and partly in granite, it was found in the same vein with native copper, and some of its ores, as the vitreous, red oxide, and arseniate of copper; also with arsenical pyrites, quartz, and fluate of lime ; but the arseniate of iron was generally met with nearer to the surface than these substances. The arseniate of iron consists of 48 per cent of iron, 18 of arsenic acid, and 32 of water.

MANGANESE.

Manganese is so difficult to be obtained in the metallic state, that very little more is known of it than that it is of a whitish or iron grey colour, somewhat malleable, brittle, almost infusible, and that its specific gravity is nearly 7.

From the black oxide of manganese, all the oxygen gas used by the chemist is obtained, and all the oxygen entering into the composition of the oxymuriatic acid consumed in the bleacheries of Britain, France, and Germany. The violet colour employed in porcelain painting is obtained from manganese. In glass-making, it is employed in the finer kinds of glass, both as a colouring material and a destroyer of colour : this application of it is ancient; it is mentioned by Pliny.

Manganese belongs perhaps rather more to primitive than to secondary countries. In the state of an oxide it is found combined in a very considerable number both of earthy and metalliferous minerals, though for the most part, only in very small proportions. It may be said to be found almost universally : it is met with both in the animal and vegetable kingdoms, and is an ingredient of mica, which is a constituent of the oldest of the primitive rocks.

Manganese is found in ores of great variety of aspect. They may be divided into four kinds.

GREY MANGANESE.

Grey manganese is either compact, acicular, or crystallized. When compact, it occurs in masses of various shapes, which are internally of a glistening metallic lustre; it is met with in France, Saxony, Piedmont, Bohemia, and at Upton Pyne in Devonshire, where it is called *Black-wad.* It has the singular property of inflaming spontaneously when mixed with one-fourth of its weight of linseed oil and set in a dry warm place. It is composed of about 83 parts of oxide of manganese, 14 of barytes, together with some silex and carbon: its specific gravity is about 4.8.

The crystallized variety occurs in acicular, rhomboidal prisms, variously placed in regard to each other, but generally in radii. Haüy has described four varieties of their crystalline forms. It consists of the oxide of manganese, with an additional but variable proportion of oxygen. It is found in nearly the same places as the former variety, and at Mendip in Somersetshire, and near Aberdeen in Scotland. Both varieties are found in primitive, secondary, and alluvial countries.

SULPHURET OF MANGANESE.

The Sulphuret of Manganese is of a blackish or brownish grey colour, and has a shining metallic lustre when newly fractured. It occurs in mass; and is fine grained, and somewhat lamellar. It is met with at Nagyag in Transylvania, in the veins in which the ores of gold and tellurium are found, for which it serves as a gangue or matrix; it also occurs in Cornwall. It consists of 82 parts of oxide of manganese, 11 of sulphur, and 5 of carbonic acid : its specific gravity is about 4.

WHITE MANGANESE.

White Manganese is of a whitish, yellowish, or pale rose red colour; and is found in the same place, and under the same circumstance, as the preceding variety. It also occurs at Kapnic and Offenbanya, in masses of various shapes. It consists of 47 parts protoxide of manganese, 40 of silex, 4.6 of oxide of iron, and 1.5 of lime : its specific gravity is about 2.8.

PHOSPHATE OF MANGANESE.

The Phosphate of Manganese is a rare mineral, having hitherto been found only near Limoges in France, in granite, and in the same vein with beryls. It is of a redish brown colour ; is hard enough to scratch glass, and may be broken into rectangular prisms with square bases. Its specific gravity is 3.46 ; and it consists of 42 parts oxide of manganese, 31 of oxide of iron, and 27 of phosphoric acid.

MOLYBDENA.

Molybdena is a rare metal, which has never been found pure : it is with difficulty reduced to a pure state, and has only been obtained in brittle infusible grains of a greyish white. It is found combined in the metallic state with sulphur; and in the acid state with lead, in the state of an oxide, forming a mineral called Molybdate of Lead, described under that head.

SULPHURET OF MOLYBDENA.

Sulphuret of Molybdena is nearly of the colour of fresh cut metallic lead. It is found massive, and disseminated, more rarely crystallized in six-sided prisms. It is opake, stains paper or the fingers, is very soft, laminated, and easily divisible in the direction of its laminæ; and is unctuous to the touch, and flexible : it yields by analysis 60 of metallic molybdena and 40 of sulphur.

It belongs exclusively to primitive countries, and is rarely found except in granite, in which it is sometimes disseminated, and therefore occasionally forms one of its constituent parts. It has been found in veins producing tin; and is generally accompanied by wolfram, quartz, and mica; less frequently by native arsenic, fluate of lime, topaz, &c.

It is found disseminated in a grey granite at the foot of a rock called Talèfre near Mont Blanc. It occurs in the tin mines of Bohemia, in the Vosges, and Sweden : near Norberg, in the latter place, in a white steatite : in Iceland, in granite, of which the felspar is red. It has lately been discovered in Huel Gorland and some other mines of Cornwall; it occurs also at Coldbeck in Cumberland, at Shap in Westmoreland, and at Glen Elg in Invernessshire.

TIN.

In its pure state, the specific gravity of Tin is about 7 ; but it has never been found pure. It is of a white colour, approaching that of silver; and is harder, more ductile, and more tenacious than lead; it is very fusible, and gives out a peculiar crackling noise while bending. It is the lightest of the ductile metals. Tin was for a considerable time supposed to have been met with in the native or pure state; but it has been pretty well ascertained that the specimens which gave rise to the opinion, were found on the sites of old smelting works, whence these specimens have since obtained the name of *Jew's-House Tin*. In the natural state, Tin is found as a nearly pure oxide, or combined, in that state, with small portions of oxide of iron and silex; it generally occurs crystallized; rarely in mass; sometimes in detached rounded pieces from the size of a grain of sand to that of a man's fist. It is also found in combination with copper, sulphur, and iron. Tin belongs exclusively to primitive countries.

The alloys of tin with other metals are mentioned in treating of lead, copper, and quicksilver. Another will be noticed under the article bismuth. In

a fine leaf, as tin foil, it is used for many purposes ;
its salts are used in dyeing : its economical purposes are well known.

OXIDE OF TIN.

Oxide of Tin rather resembles an earthy than a
metalliferous substance. It occurs in nearly colourless and translucent crystals, and in crystals of
various shades of yellow, brown, and black, which
are either translucent or opake. The form of the
primitive crystal, is an obtuse octohedron of 112°.10′
and 67°.50′, which has not hitherto been found
uncombined with the planes of one or more of the
several modifications to which it is subject : the
crystals in my possession afford about 180 varieties
of form, besides numerous compound crystals or
macles.

Tin is by no means a universally diffused metal,
many countries are entirely without it : but it is
found in Gallicia in Spain, and lately 2 or 3 veins
have been discovered in a granite mountain in Brittany in France : it occurs also at Seiffen, at Geir,
and at Altenberg in Saxony; at Schlackenwald in
Bohemia, in Banca and Malacca in the East Indies,
and in Chili in South America : but most abundantly in the western parts of Devonshire and in
Cornwall. Though Tin is the lightest of the ductile
metals, it is remarkable that the natural oxide is
one of the heaviest of the metalliferous ores : its
specific gravity is nearly 7.

The oxide of Tin belongs chiefly, and almost exclusively, to the oldest of the primitive mountains,
and is found in veins or beds, mostly the former,
in granite, gneiss, and micaceous and argillaceous
schistus.   It is often found disseminated in granite.
In veins and beds, it is accompanied by quartz,
mica, lithomarga, talc, steatite, fluate and phosphate of lime, topaz, wolfram, arsenical pyrites,
&c. which, like Tin, are considered to be among
the substances of the most ancient formation.   But
it is said rarely to be found with carbonate of lime,
sulphate of barytes, lead, or silver, which often
accompany other metals.   It is remarkable that in
the veins of Cornwall. Tin is frequently found
nearer the surface than copper.

In some of the valleys and low grounds of Cornwall, the oxide of Tin is found in grains and masses,
rounded by attrition, which frequently bear the
marks of crystallization.   The tin is generally intermixed with, or covered by, the rubbish resulting
from the disintegration of the rocks, which doubtless once held it in its native place.   Small grains
of gold are occasionally found with it.   As in order
to separate the tin from the alluvial matter, streams
of water are passed over them, these deposites are
called *Stream-works ;* one of the most remarkable
of which is in a branch of Falmouth harbour.

In these stream-works also a variety of the oxide
of Tin is found, which has obtained the name of
Wood Tin ; which occurs sometimes in wedge-
shaped pieces, banded with various shades of
brown ; and which, from their diverging and fibrous

structure, appear to be portions of globular masses; they are mostly rounded by attrition. Minute portions of this variety have lately been met with in cellular quartz : they have a very silky lustre.

TIN PYRITES.   BELL-METAL ORE.   SULPHURET OF
TIN.

This rare substance has only been found in the mine called Huel Rock in Cornwall, in a vein 9 feet wide, accompanied by sulphuret of zinc and of iron. Its colour is steel grey, passing into yellowish white : it has a metallic lustre and granular fracture, and yields easily to the knife. Its specific gravity is 4.3 ; and it consists of 34 parts of tin, 86 of copper, 25 of sulphur, and 2 of iron; but it does not seem to be ascertained in what manner the sulphur is combined with either of those metals.

TUNGSTEN.

Tungsten, called *Scheelin* by the French chemists and mineralogists, in honor of its discoverer, Scheele, is a hard, brittle, granular metal, of a light steel-grey colour and brilliant metallic lustre. It is not found in the pure state; but only in the state of an oxide, principally combined with lime, forming a mineral, commonly, though improperly, called Tungstate of Lime; or in that of an acid, combined chiefly with iron; the latter combination is called Tungstate of Iron or Wolfram.

The ores of Tungsten are chiefly, if not exclusively, found in primitive countries.

TUNGSTATE OF LIME.

Tungstate of Lime completely resembles a stone; it is commonly translucent, limpid, and of a yellowish colour: it has a laminated structure; it considerably resembles carbonate of lead, oxide of tin, and sulphate of barytes. It occurs both in mass and crystallized. The primitive form of its crystals is that of the rectangular parallelopiped; it is more commonly found in the form of an octohedron. The varieties of form assumed by its crystals are not numerous.

It is commonly found in tin veins. It occurs in those of Saxony, Bohemia, Sweden and England; and is accompanied by wolfram, quartz, mica, &c.

The translucent crystals of this mineral are composed of oxide of tungsten 78 parts; of lime 18; and of silex 3: those of Cornwall contain a little iron and manganese.

TUNGSTATE OF IRON. WOLFRAM.

Wolfram is generally of a brownish black colour, and principally occurs in veins in primitive mountains, accompanying the oxide of tin; it somewhat resembles certain ores of iron, but is generally heavier; it is met with in mass, and crystallized. It may be cleaved into a rectangular parallelopiped, which therefore is considered to be the form of its primitive crystal; the varieties assumed by it do not exceed three or four.

It is met with in the tin veins of Saxony and Bohemia, at Puy les Mines in France, and abundantly in several of the tin veins of Cornwall. In the mine called Huel Fanny, near Redruth, Wolfram is met with in the form of the primitive crystal, but very minute.

By analysis Wolfram yields 67 parts of tungstic acid, 18 of oxide of iron, and about 7 parts of oxide of manganese.

TITANIUM.

Titanium is so difficult of fusion, that the attempts to reduce it to a pure metallic state have scarcely succeeded. It is of a copper red colour.

Two of its ores are nearly pure oxides; in the others, Titanium, in the state of an oxide, is in combination with other metallic oxides, or with lime and silex.

Titanium belongs exclusively to primitive countries.

The only use to which titanium has ever been put, was in the porcelain manufactory at Sevres, where it was employed to produce the rich browns in painting it. The want of uniformity in colour occasioned its disuse.

TITANITE. RUTIL. RED SCHORL.

This mineral is of a brownish red colour, and is mostly opake, but occasionally is somewhat translucent, and is of about the hardness of quartz: it may be broken into a square prism with square bases, which therefore is the form of its primitive crystal. It is nearly a pure oxide of titanium.

It is generally found imbedded in quartz, sometimes in granite ; the hair-like appearances traversing some crystals of quartz in every direction, are generally crystals of titanite. It is found almost exclusively in primitive countries, or in alluvial deposites in their neighbourhood.

It occurs in a schistose mountain near Mont Blanc, accompanied by carbonate of lime, and some ores of iron ; at Rosenau in Hungary, and in New Castille in Spain, it is implanted in rock crystal in mountains of gneiss ; at St. Gothard, it occurs in the cavities of granite, mixed with rock crystal, &c.: and near St. Yrieix in France, and in South Carolina in America, it is found in alluvial soil. It is met with also near Beddgelu in Caernarvonshire ; and Cairgorm, and Craig Cailleach near Killin in Scotland.

ANATASE.   OCTOHEDRITE.

Anatase is also nearly a pure oxide of titanium ; it is found in octohedrons which are somewhat acute ; Haüy has described four varieties of form to which it is subject : its colour is generally bluish or redish brown. It is met with in the neighbourhood of Passau in Bavaria, at Bouen in Norway, at St. Gothard in Switzerland, and in the valley of Oysans in France, mixed with portions of granite.

NIGRINE.

In the Nigrine, the oxide of titanium is combined with about 14 of oxide of iron and 2 of manganese. It is found in primitive rocks, often imbedded in them, in Bavaria, Norway, in Mont Blanc, Mont Rosa, and in the granite of Egypt.

RUTILITE. SPHENE.

The Rutilite is composed of nearly equal parts of oxide of titanium, silex, and lime: it occurs in small crystals of a yellowish or blackish brown colour, in the form of rhomboidal prisms terminated by 4 sided pyramids: it is also found in mass. It is met with nearly in the same places as the Nigrine.

MENACCANITE.

In the Menaccanite, which is found in grains of a bluish black colour mixed with quartzose sand, in the bed of a rivulet at Menaccan in Cornwall, the oxide of titanium is combined with 54 parts of oxide of iron, a trace of oxide of manganese, and 3 of silex.

ISERINE.

In the Iserine, about 48 parts of oxide of titanium are combined with 48 of oxide of iron, and 4 oxide of uranium. It is found in angular masses and rolled pieces, near the source of the river Iser in the Reisengebirge, in granite sand; and in the bed of the river Don in Aberdeenshire in Scotland.

CEPIUM.

Cerium, in its metallic state is scarcely known; Vauquelin succeeded in procuring, by the reduction of one of its ores, a metallic globule, not larger than a pin's head. It was harder, whiter, much more brittle, and more scaly in it fracture, than pure cast iron.

Cerium has hitherto been found entering into combination in a few rare minerals, the other ingredients of which are principally earthy substances. The Cerite and the Allanite differ considerably in their composition; in both the Cerium is in the state of an oxide.

CERITE.

The Cerite was brought from the copper mine of Bastnaes, near Riddachyta in Sweden; the mine is situated in gneiss, and the Cerite was accompanied by the ores of copper, molybdena, and bismuth; and with mica, hornblende, &c.

It is generally of a pale rose colour, but sometimes inclines to brown. It occurs massive or disseminated; it is granular, brittle, and easily scratches glass; and consists of 54.5 parts of oxide of cerium, 34.5 of silex, 1.25 of lime, 3.5 of oxide of iron, and 5 of water.

ALLANITE.

The Allanite occurs in oblique four-sided, or compressed six-sided, prisms, of 117° & 63° terminated by
4 sided summits : it also occurs massive, or disseminated in black mica, and felspar. Externally it is of a
dull brownish black colour, and is opake and brittle.
It was brought from Greenland, but nothing is
known of its geological history; it obtained its
name in honour of its discoverer M. Allan of Edinburgh.   It is composed of 33.9 parts of oxide of
cerium, 25.4 of oxide of iron, 35.4 of silex, 9.2 of
lime, 4.1 of alumine, and 4 of moisture.

Two minerals, one from Bastnaes in Sweden, the
other from the Mysore, have lately been found
among others brought from those countries, which
have been found to contain cerium ; the former has
been analyzed by Berzelius, the latter by Dr. Wollaston ; they considerably resemble the Allanite in
composition.   That from Bastnaes is called the
Cerin.

URANIUM.

Uranium is a brittle, granular, hard metal, of extremely difficult fusibility.

It is remarkable that this metal has never been found in any state having a metallic appearance; consequently never in the pure state.

It is of a dark grey colour, may be cut by the knife, and is, next to tellurium, the lightest of the metals; its specific gravity being very little more than 6.

Its ores are only two in number: in both, it occurs in the state of an oxide: they are considered to belong to primitive countries. No use has hitherto been made of Uranium.

URANITE.

Uranite occurs principally in small crystals of various shades of yellow, green, and brown, which are sometimes transparent, sometimes opake. It is met with in nearly the same places and under the same circumstances as the uran-ochre. In France, at Chanteloube and St. Symphorien, and in two or three mines in Cornwall it is found in a friable granite. It is nearly a pure oxide of uranium.

At first sight the Uranite considerably resembles a variety of the arseniate of copper; but differs in the form of the crystal. The primitive crystal of the Uranite seems to approach the cube. Its crystals present several modifications. I possess 47 varieties of form, which are all from Cornwall. The variety from Gunnis Lake mine near Callington in that county exhibits quadrangular plates, very thin, of a beautiful green colour, and transparent.

The oxide of uranium is seldom found entering into the composition of other metalliferous substances, but is met with in a small quantity in the iserine, one of the ores of titanium.

URAN-OCHRE. PITCH-BLENDE.

Uran-ochre is mostly of a brown or brownish black colour; it occurs globular or massive, is sometimes disseminated, or pulverulent: it frequently resembles pitch, and is very brittle. It consists of 86.5 parts of oxide of uranium, combined with 6 of galena, 2.5 black oxide of iron, and 5 of silex.

It is met with in veins of copper, lead, silver, &c. in Bohemia, Saxony; and in two or three of the copper veins of Cornwall, passing through a friable granite.

TANTALIUM.

Tantalium is a metal, having but a slight external metallic lustre; it is dull and almost black internally; its specific gravity is little more than 6.

It is only found in the state of an oxide combined with other substances; its ores are only two. They have been found only in a primitive mountain.

TANTALITE.

The Tantalite is found principally in crystals in the form of an acute octohedron, and of a bluish grey or iron black colour. It occurs disseminated, in globular masses, in a vein composed of red felspar, traversing a mountain of gneiss, near Brokaern in Abo in Finland. It has a metallic lustre when broken; and is composed of 83 parts of oxide of tantalium, 12 of oxide of iron, and 8 of oxide of manganese: its specific gravity is 7.9.

YTTROTANTALITE.

The other compound mineral in which tantalium is found is called the Yttrotantalite, from its also containing a portion of the rare earth Yttria. It is found at Ytterby in Sweden, in a vein of felspar with the gadolinite; it occurs disseminated in masses about the size of a nut. It is nearly black; when broken, it is of a shining metallic lustre, and granular. The Yttrotantalite consists of 45 parts oxide of tantalium, and 55 of yttria and oxide of iron: its specific gravity is 5.1.

CHROME.

Chrome is a metal of a greyish-white colour, and extremely brittle : it is remarkable that it has never been found in the metallic form, either pure, or combined with any other substance, but only in the acid state, or in that of an oxide.

The chromic acid is found in combination with lead, forming a compound mineral called chromate of lead, already described.

The Chromic acid enters into the composition of the spinelle ruby.

The Oxide of Chrome is found in combination with iron ; forming a compound already described as chromate of iron. It is also found in the emerald, and in two or three other earthy minerals ; it has likewise been detected in some of the meteoric stones, or aerolites.

Chrome, as obtained in the metallic state by the chemist, from either of the foregoing compounds, has not been applied to any important use : it tinges glass of a green colour. It has been ascertained that the emerald owes its beautiful green colour to oxide of chrome : it seems therefore probable that chrome may hereafter be employed as paint.

BISMUTH.

Bismuth is of a redish-white colour, and very brittle. Its specific gravity is nearly 10.

It is found in the pure or native state somewhat alloyed by arsenic.

The ores of Bismuth are few; in one of them it is combined with sulphur; in another with inferior portions of other metals, and with sulphur. It is also met with combined with oxygen. All its ores are considered to belong exclusively to primitive mountains.

Bismuth is very little used, but it enters into the composition of some of the soft solders, and of sympathetic ink. It forms alloys with other metals. Tin and bismuth are two of the most fusible metals. The fusible-metal of Sir Isaac Newton, is composed of 8 parts of bismuth, 5 of lead, and 3 of tin; when this is thrown into water and heat applied, it melts a little before the water has reached the boiling point.

NATIVE BISMUTH.

Native Bismuth is of a silvery white, tinged with red; and occurs massive, dendritical, and crystallized in the regular octohedron, which is considered to be the form of its primitive crystal; also in cubes, and in the form of an acute rhomboid. It is rarely quite pure, but mostly contains a small portion of cobalt or arsenic : it is sometimes so disseminated throughout its gangue or matrix, as to be scarcely perceptible ; but on subjecting it to heat, globules 'of Bismuth appear on the surface. Its specific gravity is about 9.

It is met with in Bohemia, Saxony, France, Swabia, Transylvania, Sweden, and in Cornwall ; it chiefly occurs in veins in primitive mountains in a gangue of quartz, calcareous spar, sulphate of barytes, indurated clay, or of jasper, and is accompanied by ores of cobalt and nickel, and sometimes of silver, zinc, and lead.

SULPHURET OF BISMUTH.

Sulphuret of Bismuth is of a colour between tin white and lead grey, and is found massive and acicular; it is splendent or shining, and brittle. It consists of 60 parts of bismuth, 36 of sulphur, and a little iron ; but in some specimens, the proportion of sulphur does not amount to more than 5 per cent. Its specific gravity is about 6.

It is a rare mineral; but has been found at Joachimsthal in Bohemia, at Schneeburg in Saxony, and at Bastnaes in Sweden, in a gangue of quartz ; in spathose iron ore, at Biber in Hesse ; and in Cornwall.

It resembles sulphuret of antimony in colour and is liable to be mistaken for it.

A variety, of a dark steel grey colour, has been found to consist of about 43 parts of bismuth, 24 of lead, 12 of copper, 1 of nickel, 1 of tellurium, and 11 of sulphur, and has therefore been termed *Plumbo-cupriferous sulphuret of Bismuth.*

Another variety of a steel grey colour has yielded about 47 parts of bismuth, 35 of copper, and 13 of sulphur, and has received the name of *Cupriferous sulphuret of Bismuth.*

BISMUTH OCHRE.   OXIDE OF BISMUTH.

Bismuth ochre occurs both massive and pulverulent, and is of yellowish grey colour, tinged with green. It is readily reduced on charcoal to the metallic state, and is therefore considered to be a pure oxide of bismuth : its specific gravity is 4.37.

It is is very rare ; and has been principally found at Schneeburg in Saxony, accompanying native bismuth. It has also been met with in Cornwall.

ARSENIC.

Arsenic, when pure, is of a bluish white colour;
but, by exposure to air, becomes at length almost
black. Its specific gravity is above 8. It is extremely brittle, and has a granular fracture.

Arsenic is a metal of very frequent occurrence:
it is found nearly pure, when it is called Native
Arsenic, and in combination with most other metals:
its presence, when in considerable quantity, may
be detected by exposing the substance to heat, or
by striking it with a hammer, which cause the
arsenic to give out an odour like that of garlic. It is
also found in combination with oxygen; with sulphur; and in the state of an acid, with some of the
metals, and also with lime.

Arsenic belongs chiefly to primitive countries.

NATIVE ARSENIC.

Native Arsenic is found only in veins in primitive
mountains. It is of a greyish white colour and metallic lustre; but by exposure becomes dull; it
occurs in irregular masses: it is nearly pure, being
alloyed only by a very small proportion of iron, or
of gold or silver. Native arsenic is accompanied by

some of the ores of silver, cobalt, lead, nickel and iron; also by carbonate and fluate of lime, quartz, and some other substances. It is usually found in masses, somewhat round : never crystallized.

It occurs in the mines of St. Marie aux Mines in France, in those of Freyberg in Saxony; and of Bohemia, Cornwall, and Siberia.

OXIDE OF ARSENIC.

Arsenic in the state of an oxide occurs in the prismatic, acicular, and pulverulent form, in the mines of Hesse, Saxony, Hungary, and in a cobalt mine in the Spanish Pyrenees. It is also found as an efflorescence in the fissures of the lavas in some volcanic mountains.

REALGAR. ORPIMENT. SULPHURET OF ARSENIC.

Arsenic in the metallic state, combined with sulphur, forming sulphuret of Arsenic, is termed, when of a red colour, Realgar; when yellow, Orpiment.

Realgar is of a red colour, passing into scarlet, or orange. It is found disseminated, in mass, or crystallized. Its primitive crystal is the same as that of sulphur, an acute octohedron. Haüy has mentioned six varieties of its crystals. It is very tender and brittle.

It occurs in the primitive mountains of Germany,

Switzerland, Hungary, Saxony, and Transylvania. De Born mentions a vein of it between Galicia and Transylvania, about twelve feet thick. It occurs also in the vicinity of volcanoes, as of Etna, Vesuvius, &c.

It is employed as a paint; and in Siberia, it is given as a medicine in intermittent fevers.

Orpiment is of a bright lemon or golden yellow colour; it is found disseminated; in mass; or crystallized in octohedrons, which are not well defined. It seems rather to belong to stratified or secondary mountains than primitive; and is sometimes accompanied by realgar.

It occurs at Moldava in Hungary, in a vein of pyritous copper; and in a ferruginous clay at Thajoba: it is also found in Transylvania, Georgia, Wallachia, and Natolia.

It is employed as a paint. The Romans used the bright gold coloured orpiment from Syria for that purpose, and esteemed it highly.

Realgar consists of 75 of arsenic, and 25 of sulphur: Orpiment, 57 of arsenic, and 43 of sulphur.

MISPICKEL. ARSENICAL PYRITES.

This substance is of a silvery or yellowish white, and occurs in mass, disseminated, or crystallized, in almost all metalliferous primitive mountains; and abounds in many of the tin and copper veins of Cornwall. A specimen analyzed by Thomson yielded 48.1 of arsenic, 36.5 of iron, and 15.4 of sulphur;

another analyzed by Berzelius yielded only arsenic, and iron.

A variety containing from one to ten per cent. of silver is found only at Freyberg in Saxony.

The primitive crystal of Mispickel is considered to be a right rhomboidal prism. It is subject to several modifications : the crystals in my possession exhibit the planes of 7 modifications, combined in 34 varieties of form.

COBALT.

Cobalt is of a grey colour, with a tinge of red, and has the magnetic properties of iron : it is very difficult of fusion : its specific gravity is about 8.

It has never been found in the pure, or native state; but is mostly combined with arsenic and sulphur; sometimes mineralized by the sulphuric acid, &c.

The ores of cobalt occur in veins both in primitive and in secondary mountains : mostly accompanied by some of the numerous ores of copper, sometimes by native bismuth, native silver, native arsenic, and the ores of silver.

In Cornwall, cobalt occasionally occurs in copper veins; sometimes in those of a contrary direction. In one of the latter description, it is found in a mine called Huel Sparnon, near Redruth, (which is situated in argillaceous schistus) combined with bismuth, nickel, arsenic and sulphur : a block, principally consisting of these substances, which weighed 1333 lbs. was lately raised from that mine.

Cobalt is very little used except in the arts. It is brought to this country reduced to the state of an oxide, of an intense blue colour, called *zaffre*, which when melted with 3 parts of sand and 1 of potash, forms a blue glass, and when pounded very fine is called *smalts*, and is then employed to give a blue tint to writing papers, and in the preparation of cloths, laces, linens, muslins, &c.; for colouring glass, and for painting blues on porcelain. So intense is the blue of zaffre, that one grain will give a full blue to 240 grains of glass.

GREY COBALT.

The *Bright White Cobalt* of Aikin is commonly called Grey Cobalt. It is of a nearly silver white colour, but has a slight redish tinge: it occurs crystallized; yields with difficulty to the knife; is brittle; attracts the magnetic needle; gives a spark by the hammer, and yields a garlicky odour. It occurs in masses of various shapes, and in crystals of great regularity. The form of the primitive crystal is considered to be the cube; the crystals in my possession exhibit the planes of four modifications in 22 varieties of form, remarkably resembling those of the sulphuret of iron.

It consists of 44 parts of cobalt, 55.5 of arsenic, 0.5 of sulphur: its specific gravity about 6.4.

It is found in Norway; at Tunaberg in Sweden; Annaberg in Saxony; also, though rarely, in Swabia and Stiria. In Saxony and Norway, it is contained in beds of micaceous schistus, and is accompanied by quartz, pyrites, &c.

The Grey Cobalt of Aikin, and which is of a

steel grey colour, hard and brittle, is found in several of the copper veins of Cornwall : it is generally
compact and massive, and has much the aspect of
native arsenic. It consists of 20 parts of cobalt,
24 of iron, 33 of arsenic, together with some bismuth and earth, and appears to be a variety of
arsenical cobalt.

ARSENICAL COBALT.

This mineral is of a shining white colour, and is
found in masses of various forms, and crystallized
in the form of the cube : Haüy describes four varieties in the form of its crystals, which pass into
the octohedron. It does not attract the magnet.
It is found in some of the copper veins of Cornwall;
also in France and Spain ; at Annaberg and Schneeberg in Saxony ; and in Bohemia, &c. Klaproth
says that it contains arsenic, iron, and sometimes
silver and nickel. It is heavier than grey cobalt;
its specific gravity being 7.7.

EARTHY COBALT.

Earthy Cobalt is of various shades of yellow,
brown, and black. It has no metallic splendour; it
sometimes occurs in masses, sometimes almost pulverulent, and is remarkably lighter than the preceding variety, not being equal to half its weight.
It has not been analyzed, but the Cobalt is considered to be in the state of an oxide in this variety.

It is found in some of the Cornish mines; and at Alderley Edge in Cheshire in red sandstone. It is also found in Saxony, at Schneeberg and Kamsdorf; and in the Tyrol, Thuringia, &c.

RED COBALT.

Red Cobalt is also called *arseniated Cobalt,* on account of its being supposed to be cobalt mineralized by the arsenic acid; but it has not been analyzed. It is also called *Cobalt Bloom,* and passes from nearly white, through peach bloom, to a crimson red colour: it is found in small quantity in silver and copper veins. It is said to have been found in Cornwall; in Stirlingshire; and near Edinburg.

RED VITRIOL. SULPHATE OF COBALT.

Red Vitriol has been found only at Hessingrund near Neusohl in Hungary, in the form of translucent stalactites of a pale rose colour, and enclosing drops of water. It consists only of cobalt mineralized by the sulphuric acid.

NICKEL.

Nickel is of a yellowish white colour; it is attractable by the magnet, though in a degree considerably less than iron; it is ductile, and nearly as malleable as silver: its specific gravity is about 9. It has never been found pure, and its ores are only two in number.

It is remarkable that nickel, which is one of the least abundant metals, has been found by analysis to enter into the composition of meteoric iron, and of all those stony substances which in various parts of Europe and America, have fallen from the atmosphere; whence they are termed meteoric stones.

The uses of nickel are not numerous; it is chiefly employed in alloys with other metals.

KUPFERNICKEL. COPPER NICKEL.

Kupfernickel is of a pale copper-red colour, and is commonly found massive; its fracture is granular, with a metallic lustre; it yields with difficulty to the knife, but is brittle. It is hard enough to give sparks by the steel, giving out an arsenical odour. It has not been analyzed; but it is ascertained that

L 4

it consists principally of nickel and arsenic, combined with sulphur, iron, cobalt, and bismuth.

It is most abundantly found at Joachimstal in
Bohemia; Schneeberg, Freyberg, and Annaberg, in
Saxony; and Andreasberg in the Hartz.   It is also
met with in Cornwall; at Allemont in France; in
Stiria; in Arragon in Spain; and at Kolywan in
Siberia.   It is met with principally in veins in pri-
mitive mountains, accompanying the ores of silver,
cobalt, and copper.

NICKEL OCHRE.

Nickel Ochre has only been met with in the pulverulent form, generally investing the preceding ore
of nickel, and sometimes the ores of cobalt.   It has
been detected in the chrysoprase (to which it probably imparts its green colour), and in the soft
green substance in which it is found.

It is considered to be an oxide of nickel: but
has not been analyzed.

SILVER.

Silver, when pure, is soft, opake, and flexible; a piece one-tenth of an inch in diameter will support two hundred and seventy pounds without breaking. Its specific gravity is about 10. It is very white, shining, and malleable, and is found in the pure or native state; its ores are numerous. It occurs combined with antimony, iron, arsenic, lead, copper, bismuth, alumine, and silex; and mineralized by sulphur, and by the carbonic, sulphuric, and muriatic acids.

The ores of silver, whatever may be their composition, are principally met with in primitive rocks, but not of the oldest formation; they are also found in veins in secondary rocks; but never in alluvial deposites. Silver therefore is not regarded as being one of the most ancient metals.

The mines of Peru and Mexico furnish annually ten times more silver than all the mines of Europe united.

According to Helms, the mine of Jauricocha, in Peru, which is about three miles above the sea, contains a prodigious mass of porous brown iron-stone, half a mile long, as much broad, and about one hundred feet in depth, which is throughout interspersed with pure silver; and contains a white argillaceous vein, which is very much richer. It is asserted that Jauricocha and the mines of the district surrounding it, have yielded forty millions of dollars in a year.

The uses of silver are numerous, and for the most part obvious. In coin, silver is alloyed by one part of copper to fifteen of silver. The yellow colour, used in porcelain painting, is oxide of silver.

NATIVE SILVER.

Native silver, when pure, is white, and has a shining metallic lustre, but it is generally tarnished externally; it is soft, flexible, and malleable; it occurs massive, capillary, ramose, and crystallized in cubes and octohedrons; but as the structure of the crystals is not of that description which admits of regular cleavage, their primitive form has not been determined. It is less malleable than silver that has been melted in the furnace, on account, as it is supposed, of its being generally alloyed with small portions of other metals; as gold, copper, arsenic, and iron. A specimen assayed by Dr. Fordyce, yielded 28 per cent. of gold: when this metal is mixed with it, the colour approaches to that of pale brass; when alloyed by copper, it has a tinge of red.

Native silver has been found in rocks of almost every description; principally in the newer primitive. In the mines of Kongsberg in Norway, now almost exhausted, it was found in carbonate and

fluate of lime, &c ; at Schlangenberg in Siberia, on
sulphate of barytes: at Allemont, it is disseminated
in a ferruginous clay.   In Cornwall, it was found
in the Herland mine imbedded in a soft marl, and
accompanied by sulphuret of lead, cobalt, quartz,
&c. in a vein passing through argillaceous schistus :
this vein ran north and south, intersecting veins of
copper, which always in Cornwall run east and
west.   But native silver is found in Europe, most
plentifully in the mines of Saxony, Bohemia, and
Swabia.

ANTIMONIAL SILVER.

Antimonial Silver is of a yellowish white, has a
shining metallic lustre, and is often tarnished externally : it occurs in grains, massive, and crystallized in prismatic, but not very determinate crystals.
It consists of silver united with antimony in variable
proportion ; but the former, according to two analyses, is not less than 77 per cent of the mass.   It
is not abundant; but is met with in veins of calcareous spar and sulphate of barytes, accompanied
by native silver, sulphuret of lead, &c. at Guadalcanal in Spain, and in Swabia.

An ore consisting of about 12 parts of silver,
united with about 44 of iron, 35 of arsenic, and 4
of antimony, is called *Arsenical Antimonial silver.*

SULPHURET OF SILVER.   VITREOUS SILVER.

This mineral is of a dark metallic lead grey colour, and is often tarnished externally ; it is soft,
malleable, easily cut by the knife, and occurs of
indeterminate shapes, capillary, ramose and crystallized in the cube, octohedron and dodecahedron ;
but not admitting of regular cleavage, the primitive
form of its crystal has not been determined. It
consists of 85 parts of silver, and 15 of sulphur.
Its tenacity is so great, that Augustus king of
Poland had some medals struck of it.

It occurs in veins, mingled with other ores of silver, and accompanied by native silver, sulphate of
barytes, and sulphuret of lead, iron, copper and zinc.
It is found in the Saxon, Bohemian, Swabian, Hun-
garian and Norwegian mines, and it is said to have
been found in Cornwall : but the most brilliant
specimens are brought from Siberia, consisting of
groupes of crystals covered by capillary native silver,
2 or 3 inches in length.

RED SILVER.   RUBY SILVER.

This mineral is of a brilliant red colour, and is
frequently transparent ; it occurs dendritic, massive,
and crystallized, generally in the hexahedral prism,
which is sometimes modified ; it assumes about 40
varieties in the forms of its crystals, the primitive of
which is an obtuse rhomboid of 109° 28' and 70° 32',

according to Haüy. It is brittle, yields easily to
the knife, and consists of about 70 parts of sulphuret of silver, combined with about 29 parts of sulphuret of antimony. It is usually found in veins,
mingled with other minerals ; such as sulphuret of
lead, cobalt, native arsenic, realgar, grey copper,
spathose iron, iron pyrites, sulphuret of zinc, &c.
and is met with in all silver mines ; but principally
in those of Freyberg, St. Marie-aux-mines, and
Guadalcanal, &c.

BRITTLE SULPHURET OF SILVER. BRITTLE SILVER GLANCE.

The colour of this mineral is dark grey, passing
into iron black, and is of a bright and shining metallic
lustre externally ; it is soft and brittle, and occurs
massive, and in hexahedral prisms variously terminated, and in quadrangular tables. It consists
of 66.5 parts of silver, 10 of antimony, 12 of sulphur, 5 of iron, 0,5 of copper and arsenic, and 1 of
earthy impurities. It is met with in veins containing some other ores of silver and sulphuret of
lead, &c. in Hungary, Transylvania, Saxony, Bohemia, Peru, &c.

WHITE SILVER.

White silver is of a light lead grey colour, passing
into steel grey ; it is found massive and disseminated,
and has a metallic lustre ; it is soft and somewhat

brittle. Its specific gravity is 5.3 ; it consists of 48.06 of lead, 20.4 of silver, 7.88 of antimony, 2.25 of iron, 12.25 of sulphur, 7 of alumine, and 0.25 of silex. It has been procured from the mine Himmelfurst near Freyberg in Saxony, where it was accompanied by other ores of silver, and with antimony, brown spar, and calcareous spar. Some specimens have been met with passing into brittle sulphuret of silver ; others into plumose antimony.

BLACK SILVER.

Black silver is iron black, passing into a steel grey colour; it occurs disseminated, massive, and crystallized in tetrahedrons; it is somewhat hard, brittle, and has a shining metallic lustre. It is by some considered to be an argentiferous variety of the sulphuret of copper.

BISMUTHIC SILVER.

This mineral is of a light lead grey colour, which becomes deeper on exposure to the air : it occurs disseminated, rarely massive, and consists of 33 parts of lead, 27 of bismuth, 15 of silver, 4.3 of iron, 0.9 of copper, and 16.3 of sulphur. It has only been found in a mine in the valley of Shapbach in the Black Forest ; and was accompanied by quartz, hornstone, and copper pyrites.

CARBONATE OF SILVER.

Carbonate of silver is greyish, passing into iron black, and has a glimmering or shining metallic lustre. It is soft, somewhat brittle, and heavy; and consists of 72.5 per cent. of silver, 12 of carbonic acid, 15.5 of oxide of antimony, and a trace of copper. It was found about thirty years ago, accompanied by native silver, sulphuret of silver, and grey copper, in a vein of sulphate of barytes at Altwolfatch.

HORN SILVER. MURIATE OF SILVER.

This mineral is of a pearl grey colour; occasionally it is greenish blue or redish brown, and is remarkable for being so soft as easily to take the impression of the nail, and for its translucency. It has a waxy lustre; is fusible in the flame of a candle, and is generally found investing and massive, rarely crystallized in small cubes. The massive consists of 88.7 per cent. of muriate of silver, 6 of oxide of iron, 1.75 of alumine, and 0.25 of sulphuric acid. It has been found at Andreasburg in the Hartz, in the Mexican, Peruvian, Saxon, and Bohemian mines; those of Johngeorgenstadt formerly afforded large quantities of it. It is also met with in Hungary at Schemnitz; in France near Allemont, and in several mines in Cornwall, though not abundantly. It has been remarked that Horn silver is commonly met with near the surface in veins, and frequently with organic remains.

A variety is met with of a brownish white, but externally of a slate blue colour ; it is massive, dull, opake, and earthy ; and consists of about 33 parts of muriate of silver, combined with about 67 of alumine.   It is called *Buttermilk silver*, and is found at Andreasberg in the Hartz.

COPPER.

Copper, in its pure state, is so tenacious, that a wire one-tenth of an inch in diameter will support two hundred and ninety-nine pounds and a half without breaking : its specific gravity is about 8.

Copper is harder and more elastic than silver; and is the most sonorous of metals : in respect of fusibility it is between gold and iron. It is of a pale red colour, with a tinge of yellow. Its ores are numerous. It occurs in the pure or native state; also combined with iron, antimony, silver, arsenic, and with silex, lime, and water, and mineralized by oxygen, sulphur, and by the carbonic, muriatic, phosphoric and arsenic acids.

The greater part of the ores of Copper seem to belong chiefly, though not exclusively, to primitive countries ; and are found both in veins and in beds. Native Copper, the red oxide, the sulphuret, yellow copper, grey copper, and the arseniate, have been found principally in these countries ; the localities of the phosphate and muriate are less known ; but the variety of green carbonate, termed Malachite, is said to have been met with in every variety of country. The mines of Tunaberg in Sweden, and some others, as well as that of Ecton in Staffordshire, (which yielded the yellow copper ore) are situated in compact limestone. The mines of Cornwall are situated both in argillaceous schistus and granite.

Veins containing copper are not esteemed to be of so ancient formation as those enclosing tin ; because, when the veins meet with each other, those of tin are always traversed by those of copper ; but the ores of both these metals are often found in the same veins in Cornwall : the copper being generally beneath the tin.

Mines of Copper are largely wrought in England, Germany, Sweden, and Siberia ; those of Spain, France, Ireland, Norway, and Hungary, are much less extensive and numerous. Copper has been found in Asia, Africa, and America, in considerable abundance.

The uses of Copper in all its various states are almost endless, and only, if at all, inferior to those of iron. Alloyed with certain proportions of zinc it forms brass, pinchbeck, tinsel, and Dutch gold, in imitation of gold leaf. With a small proportion of tin, copper forms bronze or bell metal ; but if the proportion of tin amount to one-third, it forms speculum metal, used for reflecting tele scopes. With zinc and iron, it forms tutenag. In porcelain painting, the green is obtained from copper.

NATIVE COPPER.

Native Copper is of a yellowish red colour, has a metallic lustre, and is often tarnished externally of various colours; it occurs massive, capillary, dendritic, and crystallized, and is malleable and flexible. It assumes the form of the cube and of the regular octohedron; but not possessing that structure which allows of regular cleavage, either of these solids may be considered as the primitive form of its crystals, which are very numerous, but not very intelligible, on account of their extreme liability to that kind of compound structure which constitutes the macle. The crystals in my possession exhibit about 80 varieties of form, and were all brought from Cornwall; where Native Copper has occasionally been found in considerable abundance, accompanied by the red oxide, (into which it sometimes passes,) and occasionally by the green carbonate and the arseniate of Copper; and by quartz and fluate of lime. The copper veins of Cornwall are situated both in argillaceous schistus and granite. Native Copper is occasionally found disseminated in the serpentine of the Lizard point in that county.

Native Copper is rare in France; but is very abundant in some parts of the Uralian mountains in Siberia; at Herngrund in Saxony, the Hartz, at Fahlun in Sweden, and near the Copper Mine River within the arctic circle in America. That of Japan and that of Brazil, are said to contain a considerable proportion of gold. Quartz, fluate of lime, carbo-

nate of lime, and sulphate of barytes, usually accompany Native Copper; the two latter have rarely
been met with in Cornwall, and not at all accompanying Native Copper.

SULPHURET OF COPPER. GLANCE COPPER.

Sulphuret of Copper is of a lead or iron grey colour; it has a shining metallic lustre, and yields
easily to the knife. It occurs massive, and crystallized. The form of its primitive crystal is the hexahedral prism, which passes into an obtuse, and also
into an acute dodecahedron with triangular planes;
the crystals in my possession exhibit 83 varieties of
form besides 2 or 3 macles: they are all from Cornwall, where the sulphuret of copper has been abundantly found in several mines; sometimes intermixed with the yellow copper ore, and occasionally
accompanied by the succeeding variety; and by
spathose iron ore and fluate of lime. It is also found
at Llandidno in Caernarvonshire, and at Middleton
Tyas, in Yorkshire. It occurs also in Siberia,
Sweden, and Saxony; principally, as it is said, in
primitive mountains.

The crystallized consists of 81 parts of copper,
and 19 of sulphur; the massive contains rather
less copper, about the same sulphur, with about 2
per cent. of oxide of iron, and 1 of silex.

This mineral is of a tombac brown colour, and has an irridescent tarnish; it occurs massive, capillary, and crystallized : it is found in the cube, mostly with curvilinear faces, passing into the perfect octohedron. It consists of about 70- parts of copper, 19 of sulphur, and 7 of iron. It is generally found in the same countries as the preceding variety, and accompanying it.

GREY COPPER.

Grey copper is mostly of a steel grey colour ; it occurs massive and crystallized, and has a brilliant metallic lustre, it is brittle, but is much harder than the sulphuret of copper; it is found in the cube, passing into the regular octohedron, and in the dodecahedron with rhombic planes. I am not aware that it has been analyzed. I possess crystals of it in 26 varieties of form ; they are all from Cornwall.

This mineral seems mostly to be confounded with the *Fahlerz* of Werner, Cuivre gris of Haüy, which was formerly considered as a silver ore, and in which copper, iron, antimony, silver, and sulphur, enter into combination ; some varieties also yield arsenic : it crystallizes in the form of the regular tetrahedron, variously modified.

Grey Copper is found in Cornwall in the same

veins as the two preceding varieties; the Fahlerz, I believe, has not been found there ; but has been met with in the silver mines at Beeralston in De- vonshire, and in Wales. It is found also in Tran- sylvania, the Hartz, Saxony, &c.

This mineral is of various shades of yellow, and is often irridescently tarnished externally ; it occurs massive, stalactitic, and crystallized in the form of the regular tetrahedron, which is its primitive form ; its varieties are not numerous. It has a metallic lustre, and yields easily to the knife. It consists of copper united with variable proportions of iron and of sulphur : in general, the copper does not exceed 20 per cent. It is the most abundant of all the ores of copper, and is the chief ore of the Cornish mines, where it is found in veins, passing through argillaceous schistus and granite. It is met with in Derbyshire, and was abundant in the Ecton mine in Staffordshire in limestone. It is ge- nerally accompanied by quartz, iron pyrites, and sometimes by mispickel and the sulphuret of cop- per. In the mine called Huel Towan, it was ac- companied by spathose iron.

WHITE COPPER.

White Copper seems to be a variety of the preceding species, distinguished by its being of a silvery
white or pale brass yellow colour, and by its affording an arsenical odour before the blow-pipe; its
whiteness may perhaps be attributed to the arsenic
it contains: it has not been analyzed. It is not
common, but occasionally accompanies yellow copper; it is said to have been found in the mine
called Huel Gorland in Cornwall.

RUBY COPPER. RED OXIDE OF COPPER.

This beautiful mineral is of a fine crimson red
colour, and is frequently translucent; but externally, is mostly tarnished, sometimes of a metallic
grey colour; it yields easily to the knife, and is
brittle; it occurs massive, and crystallized in the
regular octohedron, which passes into an acute
rhomboid, the cube, and the dodecahedron with
rhombic planes: the crystals in my possession are
very numerous and exhibit about 70 varieties of
form; all of them are from Cornwall. The sp. gr.
of the red oxide of copper is 3.9; it consists of 91
copper and 9 of oxygen. A variety of the red oxide of copper, is met with in fine capillary crystals
which are lengthened cubes; another of a red or
redish brown colour, compact, and with an earthy
fracture, is called *Tile ore.*

This mineral is found at Moldava in Hungary; near Cologne; in the eastern part of the Uralian mountains in Siberia, accompanied by the variety of green carbonate of copper called Malachite; and has been met with in several mines in Cornwall, and in considerable abundance in those called Huel Unity and Huel Gorland, which are situated both in granite and argillaceous schistus; the vein was principally occupied by a brown ferruginous friable substance, called gossan by the miner; in the same vein, but above the red copper, which was generally accompanied by native copper and sometimes black copper, considerable quantities of arseniated copper, and arseniated iron were met with.

BLACK COPPER.

Black Copper occurs in a pulverulent form, investing some other of the ores of copper, chiefly the red oxide; it is generally considered to be an oxide of copper, but it gives out sulphureous vapours before the blowpipe.

CARBONATE OF COPPER.

Carbonate of Copper is of various shades of blue and of green. The *Blue* is chiefly of a beautiful azure blue, and is found in small globular masses, massive, earthy, and crystallized; it frequently accompanies the succeeding variety. I possess crystals of it, in about 30 varieties of form, but they are not very intelligible; they appear to be principally of that variety which is termed the section of the octohedron; which solid is esteemed to be their primitive form. It is not of abundant occurrence, but has been met with in the mining countries of Bohemia, Saxony, the Hartz, Siberia, &c.; also in Cornwall, and at Wanlock-head, and the Lead hills in Scotland. Some crystals from France very nearly approach the cube, others are rhomboidal, but not determinate.

The *Green* Carbonate of copper, or *Malachite*, is found massive, and in slender prismatic crystals or fibres, which are of a silky lustre, and aggregated in bundles, or stellated; frequently it is almost massive, with a silky fibrous texture. It does not present regular crystals. This beautiful mineral is said occasionally to accompany the greatest part of the other ores of copper. The finest specimens are brought from the Uralian mountains in Siberia: it is also met with in the copper mines of Saxony, Bohemia, the Tyrol, Hungary, &c.; and sometimes, though rarely, in Cornwall. The massive green variety consists of about 58 copper, 12 oxygen, 18 carbonic acid, and 11 of water. The blue variety consists of the same elements, varying somewhat in their respective proportions: and there is a variety of the green carbonate of copper, called *Chrysocolla*, which consists of the same substances in smaller proportions, together with about 26 per cent. of silex; it is of various shades of green and brown, and of very different degrees of hardness; it presents, when broken, a conchoidal fracture, and a resinous shining lustre. It is found accompanying the foregoing varieties; and has been met with in Cornwall, and in the vale of Newlands, near Keswick in Cumberland.

The *Turquoise*, so called because it was first brought from Turkey, is said to consist of the bone or tooth of an animal in the fossil state, penetrated by blue or green carbonate of copper. It is also brought from Persia.

EMERALD COPPER. DIOPTASE.

The Dioptase is of an emerald-green colour; and is met with crystallized only in the dodecahedron, the primitive form of which is an obtuse rhomboid. It is an extremely rare mineral, having only been found in a vein in Daouria, on the Chinese and Russian frontiers; it was accompanied by malachite copper. It is composed of about 29 parts of oxide of copper, 43 of carbonate of lime, and 28 of silex. The portion from which the analysis was made, was only four grains.

SULPHATE OF COPPER.

Sulphate of copper is of a blue colour, sometimes bluish green, and is generally translucent. It has a nauseous, bitter, metallic taste; and is found massive, stalactitical, or pulverulent, in certain copper mines; but it is not a common substance. It has been met with in the Parys mine

in Anglesea : and in various countries, in crystals
of eleven varieties of form, of which the primitive
is considered to be an oblique-angled parallelopiped. It consists of copper mineralized by the sulphuric acid.

MURIATE OF COPPER.

This rare mineral is of various shades of green,
and is met with in extremely minute octohedral
crystals, either loose, in the form of a green sand,
in Peru, or investing a dark ochreous quartz, at
Remolinos in Chili. The crystals in my possession
exhibit 13 varieties of form; the primitive, which
is a cuneiform octohedron, and seven of the most
simple varieties, were discovered among the green
sand of Peru : the remainder are from Chili. The
latter consists of 73 per cent. of oxide of copper,
10.1 of muriatic acid, and 16.9 of water.

PHOSPHATE OF COPPER.

Phosphate of copper is externally of a greyish
black, internally between emerald and verdigris
green ; it occurs in small rhomboids with curvilinear faces ; also massive, or disseminated in an
opake quartz. It is a rare mineral, having only
been found at Rheinbreidbach near Cologne, and
at Finneberg, and at Nassau-risingen. It is sometimes mingled with arseniated copper, and accompanied by carbonate of lead : it consists of about
68 parts of oxide of copper, and 31 of phosphoric
acid.

ARSENIATE OF COPPER.

Of this mineral there are several varieties.

It occurs in very *flat octohedral crystals*, which are of a grass green, deep blue, or bluish white colour; and are sometimes slightly transparent, with a vitreous lustre : this variety consists of 49 oxide of copper, 14 arsenic acid, and 35 water.

Another variety occurs in *six-sided tabular crystals*, which are transparent, and of an emerald green colour, or occasionally, though rarely, white and opake. The sides of the crystals alternately incline different ways, and are generally striated : but I possess some crystals much thicker than they are commonly found, of which the six sides are not striated but very brilliant. All the tabular crystals of this variety, ought, as I conceive, to be considered as sections of an octohedron ; in this opinion I am the more confirmed, because I also possess some crystals on which the solid angles of the octohedron are replaced by planes ; others on which the edges are replaced ; and again others, on which the planes of both these modifications are combined : none of these crystals have heretofore been described.

Another variety, by some called the *triedral arseniate*, is of a very beautiful bluish green, or deep verdigris colour, and transparent ; but as their surface is often decomposed and black, the crystals are then opake ; their form is an octohedron, which is generally elongated, and their summits are sometimes deeply replaced, giving them the appearance of six-sided prisms with diedral summits; in others, two of the four lateral edges are also deeply replaced, the crystals then assume the appearance of four-sided prisms with diedral summits. The crystals of this variety also assume the form of the tetrahedron, and of a very acute rhomboid, sometimes perfect, sometimes passing into the octohedron; they have been said to occur also in the rare form of the triedral prism, but as the acute rhomboids are often placed on the gangue on one of their extremities, having the other, which if perfect would appear as an acute apex, deeply replaced by a regular triangular plane, I conceive this appearance has given rise to the opinion that the crystals assume the form of triedral prism.

The preceding varieties differ in their respective proportions of oxide of copper, arsenic acid, and water, from the first variety: in the succeeding variety there is no water.

This variety occurs in *slightly acute octohedrons,* which are usually of a bottle green colour; sometimes brown, or nearly black, and somewhat transparent. These crystals are mostly elongated; in some of them, the summits of the octohedron are replaced, as well as two of the four lateral edges; and as these crystals are generally long, they assume the appearance of four, six, or eight-sided prisms having diedral summits, whence this variety has been termed the *prismatic arseniate.* It sometimes exhibits capillary crystals of indeterminate forms;

and others which are regular for some length, but fibrous at the extremity.

The two following varieties agree in their respective proportions of oxide of copper, arsenic acid, and water, but differ from the two first varieties.

One of them, which is of various shades of green, brown, yellow, and white, is of a fine diverging fibrous structure, and silky lustre : it is called the *Hæmatitic arseniate* of copper.

The other occurs in extremely minute, flexible fibres, occasionally so small as to have the appearance of dust; they are of various shades of blue, green, brown, yellow, and white ; and possess a silky lustre.   It is called the *Amianthiform arseniate.*

Martial Arseniate of Copper, which, until lately, has been termed *Cupreous Arseniate of Iron,* is of a pale bluish green colour, and occurs in small four, six, or eight-sided prisms, with tetrahedral summits, generally grouped in small globular radiated masses ; they are transparent, and have a shining vitreous lustre.

All the above varieties of the arseniate of copper were found in the same veins which produced the red oxide of copper, in the mines called Huel Gorland and Huel Unity, which adjoin each other in Cornwall ; their veins pass through both granite and argillaceous schistus.

GOLD.

The specific gravity of Gold, when pure and beaten, is about 19; it is very soft, and perfectly ductile and flexible. So great is its tenacity that a piece one-tenth of an inch in diameter, will hold five hundred pounds without breaking; and it is computed that a single grain of gold will cover the space of fifty-six square inches, when beaten out to its greatest extent.

Gold is always found in the metallic form, whence by mineralogists is is said to occur in the native or pure state; but it is generally alloyed by small portions of other metals, as silver, copper, &c.

The uses of gold are well known. Alloyed by copper, it is employed for ornamental purposes, coin, and plate.

In English coin it was alloyed by two parts of copper, to twenty-two of gold. The alloy in gold used in plate, was formerly the same as the coin : it is now 18 carats, or $\frac{18}{24}$ths gold. The purple colour used in porcelain painting is obtained from a preparation of gold.

NATIVE GOLD.

Native Gold, is yellow, orange yellow, or greyish yellow, with a shining metallic lustre; it occurs

crystallized, capillary, ramified, and in masses of
various sizes, from the weight of very minute portions
to that of several pounds ; it is soft, inelastic, flexible, and malleable. It is rarely perfectly pure, but
mostly contains small portions of other metals ; as
of silver, copper, &c. It occurs crystallized in the
form of the cube and octohedron, but as its crystals
do not admit of regular fracture, their structure is
not sufficiently known to enable mineralogists to
decide which of those two solids is the form of its
primitive crystal. I possess crystals exhibiting 21
varieties of form, besides 12 others in that compound
species of crystallization, expressed by the term
macle ; each consisting of equal portions of the octohedron united together.

Gold is sometimes combined in other metalliferous
minerals, in various proportions, but is said always
to be in the metallic state. An argentiferous variety
has yielded by analysis 36 per cent. of silver ; and
an auriferous variety of silver (mentioned under
that head) 28 per cent. of gold. It is combined
with other metals in the ores of tellurium, and not
unfrequently forms an ingredient in iron pyrites,
which thence is termed auriferous. It is also said
to have been occasionally found in certain sulphurets of iron, zinc, lead, and mercury, and in some
varieties of copper and of arsenical pyrites.

In veins, gold is found only in primitive mountains, but not in those of the most ancient formation ;
these veins principally contain quartz, felspar, carbonate of lime, and sulphate of barytes ; but the
gold is sometimes accompanied by sulphuret of iron,

of silver, and of lead, and occasionally by red silver, manganese, grey cobalt, and nickel.

Gold is found in veins; also in rivers, and in alluvial matter, in several countries of Europe. From Spain, the Phœnicians and Romans are supposed to have drawn their principal riches; it is also found in Germany and Sweden; and very sparingly in France and Italy. The principal European mines are those of Cremnitz and Chemnitz in Hungary; which, together with some others of inferior note, annually produce, by estimate, about 2000 pounds weight. Small quantities are also found in alluvial deposites in Switzerland, and in Ireland; the latter of which contains about 15 per cent. of silver. It is found occasionally in small grains intermixed with tin in some of the stream-works of Cornwall.

In Asia; gold is found in Siberia in veins, and in many of the Asiatic islands in sands.

From the rivers and alluvial deposites of Africa, large quantities were furnished to the ancients.

By far the greater part of the gold now brought into use, is obtained from the rivers of South America; in various parts of which continent, it is also found in veins, and in considerable abundance. In the Vice Royalty of La Plata alone there are 30 gold mines or workings. It is calculated that the annual produce of America is about 30,000 pounds weight.

Helms says, that when a projecting part of one of the highest mountains in Paraguay fell down, about thirty years ago, pieces of gold, weighing from two to fifty pounds each, were found in it.

PLATINA.

The specific gravity of Platina when pure, is about 23; its colour is between tin-white and iron-grey. Its malleability is so considerable that it may be beaten into leaves as thin as tin foil, and its ductility so great, that Dr. Wollaston has succeeded in drawing it into a wire $\frac{1}{18750}$th part of an inch in diameter, which will support about one grain and one-third of a grain without breaking. It possesses considerable elasticity, and in hardness is not much inferior to iron; but is very difficult of fusion. It is only found in the native state.

Pure Platina in thin plates is very ductile and flexible. Of late it has been formed into mirrors for reflecting telescopes, spoons, crucibles, and some vessels of considerable dimension for the use of the chemist in particular processes.

NATIVE PLATINA.

Native Platina, is between steel-grey and silver-white colour, and is nearly as hard as iron, and malleable, but is infusible. It has hitherto only been found in small flattened grains rarely exceeding the size of a pea; the largest that has been seen is of the size of a pigeon's egg, and was presented by Humboldt to the King of Prussia. Native Platina is much lighter than pure platina: it has been found in St. Domingo, Brazil, and Peru.

In St. Domingo it is met with in the eastern part of that island, in the sands of a river called Jaki, at the foot of the mountains of Sibao. The grains are somewhat larger than those of Peru, and are accompanied by magnetic iron ore, gold, &c. It has not been analyzed.

In Brazil it is found in the gold mines of that country, in small grains of a spongy texture, free from tarnish, and with very little lustre, mixed with grains of gold and of palladium; perhaps also with the natural alloy of iridium and osmium. It does not contain any of the magnetic iron sand, or of the minute hyacinths, which always accompany the Peruvian ore. It consists of platina alloyed by very minute portions of gold and of palladium.

In Peru, it is only met with in the Rio del Pinto, in the districts of Citara and Novita in the province of Choco, and near Carthagena in New Grenada. It is found in a magnetic iron sand, in which are mixed grains of gold, minute hyacinths, and fossil wood: and it is said that the whole is covered by rounded pieces of basalt enclosing olivine and py-roxene. The grains of Platina are small, flattened, and have occasional indentations, the surfaces of which are generally tarnished; but the other parts have a shining metallic lustre. It consists of platina alloyed with small proportions of iron, copper, lead, palladium, iridium, rhodium, and osmium.

The grains of Crude Platina analyzed by Descotils were accompanied by grains of menachanite and of chromate of iron.

RHODIUM.

This metal has hitherto been found only alloying the native platina of Peru. When pure, Rhodium has a bright metallic surface, but is not malleable; its specific gravity is about 11.

IRIDIUM. OSMIUM.

The former of these two metals, when pure, is white, and perfectly infusible; the latter is of a dark grey or blue colour. They occur, alloying native platina in very small proportion; and likewise together, forming a natural alloy of the two metals.

ALLOY OF IRIDIUM AND OSMIUM.

This natural alloy is found accompanying native platina, in the form of very small, irregular, and flattened grains, which have a shining metallic lustre, but are of a somewhat paler colour than native platina, and are harder and heavier; their specific gravity being 19.5: they possess a lamellar structure, and are brittle.

PALLADIUM.

The specific gravity of Palladium, when pure, is about 11. In colour, it greatly resembles platina; in thin laminæ it is very flexible, but not very elastic; it is somewhat harder than bar-iron, and is very malleable.

It occurs, together with some other metals, alloying, in small proportion, the native platina of Brazil: and also in the native state.

NATIVE PALLADIUM.

Native Palladium occurs in grains apparently composed of diverging fibres; in other respects these grains differ little in external character from those of the native platina, amongst which they are found. Native palladium is infusible; its specific gravity is 11.8; and it consists of palladium, alloyed by minute portions of platina and iridium.

TELLURIUM.

Tellurium, when pure, is about the colour of tin; it is brittle, and nearly as fusible as lead; its specific gravity is little more than 6.

It is an extremely rare metal, and is found only in the metallic state; but is always alloyed, though in very different proportions, by other metals. Its ores are few and rare.

NATIVE TELLURIUM.

Native tellurium is of a tin white colour, and has a metallic lustre: it occurs in small grains, which are brittle, and yield to the knife. It very much resembles grey antimony. It is found in the veins of a transition mountain of compact carbonate of lime, at Fazebay in Transylvania; and also at Beresof in Siberia. It consists of 92.55 parts of tellurium, 7.2 of iron, and 0.25 of gold. Its specific gravity is about 6. It is procured for the sake of the gold it contains, though so small in quantity.

GRAPHIC TELLURIUM. AURUM GRAPHICUM.

This mineral is of a steel grey colour, with a splendent metallic lustre; and occurs in small flat

six-sided prisms, with or without four-sided summits; the crystals are generally disposed in rows
on the surface of quartz, and are so arranged as to
give the appearance of certain characters; whence
its name: it is also sometimes found in granular
masses: it yields easily to the knife, and is brittle.

It is met with only at Offenbanya in Transylvania,
together with sulphuret of zinc, pyrites, grey copper,
&c. in veins which traverse a porphyritic mountain.
Its specific gravity is 5.7; and it is composed of 60
parts of tellurium, 30 of gold, and 10 of silver.

PLUMBIFEROUS TELLURIUM.

This substance is either yellow or black. The
yellow variety occurs in grains and in minute flat
four-sided prisms, of a bright metallic lustre; it is
somewhat flexible, and soft. It consists of 44.75
parts of tellurium, 26.75 of gold, 19.5 of lead, 8.5 of
silver, and 0.5 of sulphur.

The Black is found in irregular shapes, or in
lengthened and six-sided plates of a shining metallic
lustre. It consists of 32.2 of tellurium, 54 of lead,
9 of gold, 1.3 of copper, 3 of sulphur, and 0.5 of
silver.

This variety, which is much heavier than the former, is only found at Nagyag in Transylvania, and
is procured as an ore of gold; it is accompanied
principally by the same substances as the former
variety.

ANTIMONY.

Antimony is a compact, brittle, bluish white metal, whose specific gravity is between 6 and 7; it is found nearly pure.

The ores of antimony are only five in number; all of which have not been analyzed. In some of them, it is found combined with oxide of iron, arsenic, silex, sulphur, and oxygen.

Antimony is found both in primitive and secondary countries. It forms alloys with other metals, and is used in the arts. It enters largely into the composition of printing types; it is also used in medicine.

NATIVE ANTIMONY.

This substance is found of irregular shapes; never crystallized regularly. It occurs at Sahlberg in Sweden in calcareous spar; at Allemont in Dauphiné, in white quartz; at Andreasberg in the Hartz, in quartz and spathose iron ore.

The form of its primitive crystal is the regular octohedron.

It consists of 98 parts of antimony, the rest being silver and iron; but some specimens, on being exposed to heat, give out a garlickv odour, indicating the presence of arsenic.

GREY ANTIMONY.   SULPHURET OF ANTIMONY.

Grey antimony is of a light lead grey colour externally, but presents when fractured a brilliant metallic lustre ; it is occasionally lamellar, fibrous, and is often crystallized : it is extremely brittle, and so fusible, that it readily melts in the flame of a candle : when in minute capillary crystals, it is termed *plumose antimony.*

The form of its primitive crystal has not been ascertained. It occurs in oblique four-sided prisms, terminated by four-sided pyramids ; the crystals in my possession exhibit 14 varieties of form. It consists of 75 parts of antimony and 25 of sulphur.

It is found in Saxony, Hungary, and France; also in Cornwall and Dumfrieshire. It is mostly met with in micaceous schistus, or clay porphyry, mixed with oxide of iron ; and is accompanied by quartz and spathose iron ore ; in Hungary by sulphate of barytes, sometimes calcareous spar, fluor spar, and chalcedony. It is remarkable that in Cornwall it is only met with in veins in a direction contrary to that of the copper and tin veins, which they pass through.

RED ANTIMONY.

This mineral is brownish, bluish, or redish externally ; and is principally found in minute diverging crystals, which are brittle.

It often accompanies the preceding varieties, and is met with in Hungary, Saxony, in Dauphiné, but principally in Tuscany. It consists of about 67.5 parts of antimony, 10.8 of oxygen, and 19.7 of sulphur.

WHITE ANTIMONY. OXIDE OF ANTIMONY.

White antimony is of a white, yellowish white, or grey colour; it is rarely found in mass, but mostly in slender diverging crystals, which are very tender, heavy, and translucent.

It is a rare mineral; at Allemont in France it is found on native antimony. In Bohemia, on sulphuret of lead. In Saxony, Hungary, and Spain, it is met with investing sulphuret of antimony. It consists of 86 parts of oxide of antimony, 8 parts of silex, and 3 of oxide of antimony and oxide of iron.

ANTIMONIAL OCHRE.

Antimonial ochre has been found at Endellion in Cornwall, in Saxony, and Bohemia, upon some of the ores of antimony; it is earthy, of a yellowish or brown colour, and is extremely rare.

LEAD.

Lead is of a bluish grey colour, and is malleable, ductile, inelastic, and very soft ; it has never been found in the pure or native state : its specific gravity is between 11 and 12.

The ores of lead are numerous ; they appear under very different circumstances and aspects, and present a considerable diversity of combination. Lead is found mineralized, mostly in the state of an oxide, by sulphur and by the carbonic, muriatic, phosphoric, arsenic, molybdic, and chromic acids, and by oxygen : it is also found in combination with the metals antimony, iron, manganese, and silver, or their ores; with the earths, silex, alumine, lime, and magnesia, and with water. Some of the ores of lead, which are very numerous, present combinations of several of these substances ; a few of them have a metallic aspect, but several of them have rather the appearance of earthy minerals, being in considerable degree transparent or translucent. The ores of lead chiefly occur in secondary countries ; sometimes in the veins of primitive mountains.

It would scarcely be possible to enumerate all the valuable purposes to which lead is applied in the arts, in medicine, and in the common wants of man. Among its less obvious uses, lead is employed to glaze pottery, and its oxide enters into the composition of glass. Four parts of lead and one of antimony form printing types, to which by some is added a little copper or brass. With tin and bismuth it forms alloys, which are used in the arts.

GALENA. SULPHURET OF LEAD.

Galena has nearly the colour and lustre of pure lead : it is met with crystallized in the form of the cube, which is that of its primitive crystal, and in nine varieties of form ; among which is the regular octohedron. It occurs also specular, radiated, granular, and compact. Its specific gravity is 7.5 ; and it consists of about 85.13 per cent. of lead, 13.8 of sulphur, and 0.5 of oxide of iron ; but carbonated lime and silex are found entering into combination in some varieties, in proportions varying between 29 and 38 per cent. Galena is rarely found without some proportion of silver, which varies from $\frac{1}{300}$ or less, to $\frac{1}{12}$. The presence of silver is said considerably to diminish the lustre of the Galena, and it is also said that it is much more frequently found in the octohedral, than in the cubical varieties.

The specular variety is sometimes called *Looking Glass* Lead ore, on account of its great brilliancy, and in Derbyshire, *Slikenside ;* the radiated variety is said always to contain a portion of antimony. The granular variety is sometimes nearly as fine-grained as steel.

Galena is almost the only ore of the numerous ores of lead, which is found in sufficient quantity to be wrought for the lead it contains. This substance occurs under great diversity of circumstance; which, owing to the importance of the mineral, deserves a slight notice.

In France, in the mine of Pompean, it is accompanied by fossil wood: near Medrin, it traverses nearly perpendicular beds of limestone: near Vienne it occurs in schistus; in Languedoc and the Vosges, in decomposed granite; and in some places in large veins, passing through primitive mountains.

At Bleyberg in Carinthia, it occurs in beds, alternating with beds of compact carbonate of lime; and in grains disseminated through sandstone, and accompanied by oxide of copper, and brown iron ore.

In Silesia it occurs in veins, and in rounded masses in horizontal beds of ferruginous marl, resting upon thicker beds of compact carbonate of lime, enclosing fossil shells, and asphaltum.

In Spain, the most important mines of sulphuret of lead are situated in granite hills, in the province of Jaen, and near the city of Canjagar.

In England, the most important mines are those of Derbyshire, which are principally situated in compact limestone, enclosing shells: the veins of lead ore are sometimes nearly vertical, occasionally horizontal, and they sometimes open into large

caverns. In these mountains is found the amygdaloid or toad stone, which interrupts the vertical
veins, but not the horizontal veins, or rather beds.
The lead ore is accompanied by carbonated lime,
sulphate of barytes, (of the variety called cauk)
and fluate of lime ; occasionally by petroleum and
elastic bitumen. It is confidently asserted that when
the variety of sulphuret of lead called *Slikenside*,
is met with, and by any means disturbed a terrible
explosion ensues, by which considerable masses are
detached from the vein : this singular circumstance
has not been explained.

This substance is found in almost every mineral
district in the known world, and perhaps, next to
certain ores of iron, is the most common of metalliferous ores : but it is said not to have been met
with in any considerable quantity in the Altaic or
Uralian chains of mountains in the northern parts
of Asia ; nor is it common in Peru.

BLUE LEAD ORE.

Blue Lead Ore has only been found as Zschoppau
in Saxony, in veins, accompanied by some other
ores of lead, and with quartz, fluor spar, &c. ; it
occurs massive, and crystallized in small six-sided
prisms, and is of a colour between lead grey and
indigo blue, with a slight metallic lustre.

TRIPLE SULPHURET OF LEAD.

This mineral is generally of a dark lead grey colour, and shining metallic lustre; it is mostly crystallized in the form of the cube and its varieties; it yields easily to the knife, and is very brittle. It consists of 50 parts of sulphuret of lead, 30 of sulphuret of antimony, and 20 of sulphuret of copper. It has hitherto only been found in Huel Boys mine in Cornwall, in a north and south vein passing through argillaceous schistus, and accompanied by sulphuret of zinc; but some minerals very nearly approximating to this substance in composition have been met with in other countries.

NATIVE MINIUM.

Native Minium is believed to be a pure oxide of lead, which does not appear to have been ascertained by analysis. Its ordinary colour is scarlet, but it is also met with of various shades of grey, yellow, and brown; it occurs of indeterminate shapes, and pulverulent; the latter variety is found in small beds alternating with clay and sulphuret of lead. It occurs in several places in Saxony, Germany, and in France; also at Grassington Moor, in Craven; and at Grasshill Chapel in Yorkshire.

CARBONATE OF LEAD.

This beautiful mineral is white, or of various shades of grey and brown, and of a resinous lustre. It occurs crystallized, acicular, and fibrous. The crystals are translucent or transparent; the acicular and fibrous mostly opake: it yields easily to the knife, is brittle, and possesses double refraction in a high degree. The primitive form is a rectangular octohedron; its crystals are found in 12 varieties of form. The specific gravity of Carbonate of Lead is 6.7; and it consists of 77 per cent. of lead, 5 of oxygen, 16 of carbonic acid, and about 2 of water. The Carbonate of Lead is not very abundant; it is not found in large masses, and is always accompanied by other ores of lead. It is met with in Languedoc and Brittany, in France; in the Hartz; in the lead hills in Scotland; at Alston Moor in Cumberland; in Durham; and occasionally in Cornwall, but principally of the acicular variety.

Sometimes this substance is tinged of a green colour on the surface, by the carbonate of copper; occasionally it is of a metallic lead grey, exhibiting the partial conversion of the carbonate into sulphuret of lead: and at Grassfield mine near Nent-Head in Durham, Carbonate of Lead is found abundantly of an earthy texture, and of a grey colour; but is tinged sometimes greenish, yellowish, or redish: it occurs massive, or granular, and is very heavy.

MURIATE OF LEAD.

Muriate of Lead is of a greenish yellow colour, and is found crystallized in quadrangular prisms, which are sometimes terminated by pyramids; it is soft and somewhat transparent, and consists of 85.5 of oxide of lead; 8.5 of muriatic acid, and 6 of carbonic acid. It is found at Cromford Level near Matlock, and in the mountains of Bavaria.

PHOSPHATE OF LEAD.

Phosphate of Lead is of various shades of green, yellow, and yellowish brown; but when reduced to powder is always of a grey colour. It is found principally in six-sided prisms, sometimes having six-sided pyramids, but does not afford many varieties of form; it is divisible into an obtuse rhomboid, which therefore is considered to be its primitive crystal. Its crystals are generally somewhat translucent, possess a resinous lustre, and are brittle. The green phosphate of lead consists of 80 parts of oxide of lead, 18 of phosphoric acid, and nearly 2 of muriatic acid.

The brown variety contains about 2 per cent. more of the phosphoric acid, and 2 per cent. less of the oxide of lead. A variety is found at John-georgenstadt in Saxony, consisting of about 77 parts oxide of lead, 9 of phosphoric acid, and 4 of arsenic acid, the rest being water.

N

The Phosphate of Lead occurs in veins in primitive and secondary mountains ; it sometimes accompanies sulphuret of lead, carbonate of lead,
iron ochre, quartz, sulphate of barytes, and carbonate of lime. The green phosphate occurs at
Alston Moor in Cumberland, at Allonhead, Grasshill, and Teesdale in Durham ; at Nithisdale in
Yorkshire ; and at Wanlockhead in Scotland.

SULPHATE OF LEAD.

This substance mostly occurs in translucent crystals, which are colourless, or of a smoke or yellowish grey colour. The form of the primitive crystal
is a rectangular octohedron ; the crystals in my
possession exhibit 30 varieties of form ; they are all
from Cornwall. Sulphate of Lead is composed of
about 71 parts of oxide of lead, 24.8 of sulphuric
acid, 2 of water, and 1 of oxide of iron. It has
been found in Andalusia in Spain ; at Wanlockhead
and the Lead hills in Scotland ; but principally in
the Parys mine in Anglesey. In Cornwall, it was
met with in a copper vein in a mine called Veleno-
weth, very near the surface, and was accompanied
by the Sulphuret of Lead ; it occurred in an ochreous brittle substance, termed by the miner Gossan.

ARSENIATE OF LEAD.

The Arseniate of Lead occurs principally in slender six-sided crystals, which sometimes are fasciculated ; or in fibres, of various shades of yellow, sometimes with a tinge of green ; they are generally translucent, and have a resinous lustre. The specific gravity of Arseniate of Lead is about 6 ; it consists of 69.76 per cent. of oxide of lead, 26.4 of arsenic acid, and 1.58 of muriatic acid. In France, it has been met with in a lead mine, accompanied by quartz, fluate of lime, and sulphuret of lead. In Andalusia, in felspar, with quartz and galena ; and in Huel Unity mine in Cornwall, in a copper vein situate in granite.

MOLYBDATE OF LEAD.

The Molybdate of Lead is met with principally in crystals of various shades of yellow, having a glistening resinous lustre ; it is soft, brittle, and somewhat translucent. The primitive crystal is an octohedron, with similar and equal isosceles triangular planes. The crystals in my possession exhibit 35 varieties of form. The specific gravity of Molybdate of Lead is about 5 ; and it consists of 58.4 parts of oxide of lead, 38 of molybdic acid, and 2.08 of oxide of iron. It was first discovered at Bleyberg in Carinthia, upon a compact limestone ; and has been since found at Zimapan in Mexico, on the same substance. It occurs also near Freyberg and at Annaberg in Saxony, and at Felsobanya in Hungary.

CHROMATE OF LEAD.

This beautiful substance is of an orange red colour ; it has mostly been met with crystallized : the primitive form of its crystals is an oblique four-sided prism ; the varieties it assumes are very few. It consists of 64 parts of oxide of lead, and 36 of chromic acid. This mineral is extremely rare ; it was found in the gold mine of Beresof, in the Uralian mountains in Siberia, upon a quartzose gangue containing oxide of lead and oxide of antimony, which occurred in a vein containing sulphuret of lead, parallel with another containing decomposed auriferous pyrites. These veins are situated in gneiss and micaceous schistus. Pallas mentions having also discovered this mineral 15 leagues higher north, disseminated in beds of clay, and dispersed on beds of sandstone, alternating with each other, and accompanied by cubic crystals of auriferous pyrites.

This substance is said to be occasionally accompanied by small acicular crystals of a green colour, which are considered to consist of oxide of lead and oxide of chrome, but have not been analyzed.

ZINC.

Zinc is a bluish grey metal; its tenacity is not great; a piece one-tenth of an inch in diameter will hold twenty-six pounds without breaking; and being far less ductile than some other metals, its importance is thereby diminished. Its specific gravity is about 7.

Zinc is never found in the pure metallic state, but mineralized by sulphur, oxygen, the carbonic or the sulphuric acids; and combined with oxide of iron, silex, and with water. All the varieties of its ores may be said to be comprehended in the four following species, most of which have the appearance rather of earthy than of metalliferous substances; they belong chiefly to secondary countries.

Zinc is employed by the Chinese for coins: it enters into the composition of many alloys. It is sometimes used in medicine, and in oil painting.

BLENDE. SULPHURET OF ZINC.

Blende is met with of various shades of yellow, brown and black; it occurs of indeterminate shapes, massive and crystallized; it yields easily to the knife, and is brittle.

N 3

The form of the primitive crystal of Blende, into which the compact variety is readily reduced by cleavage, is the rhomboidal dodecahedron ; but the varieties of form assumed by its crystals are very numerous, though not very intelligible, except such as are obviously allied to the tetrahedron, octohedron and the cube.

The massive variety of a brown colour, affords 50 parts of zinc, 12 of iron, and about 29 of sulphur ; some varieties are phosphorescent by friction.

Blende is found in most mineral countries, especially in beds in the older secondary. It is met with in metalliferous veins traversing primitive mountains, principally in those containing copper and lead ; it often accompanies, or is accompanied by, iron pyrites, native silver, grey antimony, spathose iron, sulphate of barytes, calcareous spar, and quartz. It is very abundant in many of the copper and tin veins of Cornwall, especially the former, and at a small depth beneath the surface. Some of the blendes of Hungary and Transylvania are auriferous.

CALAMINE.   CARBONATE OF ZINC.

This substance is found earthy, compact, and crystallized ; it yields easily to the knife, and has remarkably the appearance of an earthy or stony substance.

The form of its primitive crystal is an obtuse rhomboid. It does not assume many varieties of form : but is often found investing crystals of carbonated lime; which, in some instances, being decomposed, leave the calamine in the forms they had assumed.

It is chiefly found accompanying sulphuret of lead, in shell limestone; and is particularly abundant in the Mendip hills in Somersetshire; at Holywell and other places in Flintshire; in Derbyshire; and in Carinthia, &c. In France, near Juliers, it forms very extensive beds, and is accompanied by certain ores of lead and iron.

It yields about 65 per cent. of oxide of zinc, and 35 of carbonic acid; a variety is found in the Rutland mine at Matlock, which is combined with carbonate of copper.

ELECTRIC CALAMINE.

The name of this mineral is derived from its property of becoming electric when gently heated. Its colour is greyish, bluish, or yellowish white; it is found in mass, and also crystallized in small flat hexahedral prisms, which are harder than common calamine.

It is found in Hungary, at Fribourg; and in Leicestershire, Flintshire, and at Wanlock-head. It differs from the other ores of zinc in always containing a considerable proportion of silex. The variety from Wanlock-head yields 66 oxide of zinc, and 33 of silex.

SULPHATE OF ZINC.

This mineral is a white, limpid, soluble salt, and is by some mineralogists ranked amongst saline minerals; it has a nauseous metallic taste, and is found filamentous, massive, and stalactitical.

In the natural state it is rare, and chiefly occurs in capillary efflorescences, or in stalactites, on the sides of the workings in veins of sulphuret of zinc. It is thus occasionally seen at Ramelsberg in Switzerland, at Idria in Carniola, and at Schemnitz in Hungary. It also occurs at Holywell in Flintshire. That of Ramelsberg yields by analysis about 27 parts of oxide of zinc, 22 of sulphuric acid, 50 of water, and a trace of oxide of manganese.

QUICKSILVER or MERCURY.

The liquidity of Mercury at the ordinary tempe-rature of the atmosphere, is a remarkable character, and distinguishes it from all other metals. It is thirteen times heavier than water. It is found pure; and also combined with silver, with sulphur, and with small quantities of silex, oxide of copper, carbon and bitumen; and mineralized in the state of an oxide, by the muriatic, and sulphuric acids.

Its ores are not numerous; and being rarely found in primitive rocks, it is not considered to be a metal of the newest formation.

The quicksilver mines of Idria, in Saxony, are said to yield 100 tons annually; and those of Spain a still greater quantity. The mines of Peru are by some supposed to be still richer.

The uses of mercury in medicine, in the arts, and in experimental philosophy are numerous; but its chief use is in the separation of gold and silver from their ores, by a process called amalgamation. When amalgamated with tin, and laid on glass, it forms mirrors.

NATIVE QUICKSILVER.

Native Quicksilver is of a silver white colour, and splendent metallic lustre; it occurs disseminated in globules, or collected in the cavities of its mines, which are commonly situated in calcareous rocks, or indurated clay, or argillaceous schistus. It is mostly met with in the mines containing the ores of quicksilver. It sometimes contains a little silver.

Quicksilver is found in the Palatinate, Saxony, Bohemia, Hungary and Transylvania; and abundantly in Peru.

NATIVE AMALGAM. SILVER AMALGAM.

This mineral is of a silver white, or of a greyish colour, and is sometimes semi-fluid; when compact it is very brittle, which at once distinguishes it from silver: it is mostly tarnished externally. It occurs also in small octohedrons, in rhomboidal dodecahedrons, and in thin laminæ; and is commonly found in a kind of clay, which is of various colours. It consists of 64 of mercury, and 36 of silver. It is a rare mineral, and has principally been met with at Rosenau in Hungary, and at Mærsfeldt and Moschellandsberg in the dutchy of Deux-ponts. It is said to be found in veins containing silver, traversing those enclosing quicksilver.

CINNABAR.

Cinnabar is of various shades of red, frequently cochineal red. It is very heavy; it occurs massive, when it is dull and opake; it also occurs of a minutely fibrous structure, with a glimmering silky lustre; also lamellar, of a shining lustre and translucent; and crystallized in the regular hexahedral prism, which is considered to be the form of its primitive crystal; only one variety of form has been noticed. It consists of 81 of mercury, 15 of sulphur, and 4 of iron.

A variety called *Hepatic Cinnabar* is united with small portions of carbon, silex and oxide of copper; and this variety sometimes occurs mixed in various proportions with coarse coal or bituminous shale, and is then called *Bituminous Cinnabar.*

The most abundant European mines of Cinnabar, are those at Idria in Carniola (which principally yield the hepatic variety), and those of Almaden in Spain, which are situated in the independent coal-formation. The ores of Cinnabar are usually accompanied by calcareous spar, spathose iron, micaceous iron, and iron and copper pyrites.

Cinnabar is said to occur sparingly in primitive strata.

HORN QUICKSILVER.

Horn Quicksilver is of a pearl grey colour, sometimes of a greenish yellow; it is soft, translucent,
and of a vitreous lustre; it occurs massive; also crystallized in small 4 sided short prismatic crystals,
terminated by 4 sided pyramids, and therefore in
dodecahedrons; but the planes of the summits are
rhombic, the lateral planes at six-sided. It consists
of about 76 parts of oxide of mercury, 16 of muriatic acid, and 7 of sulphuric acid. It is found at
Almaden in Spain, at Horsowitz in Bohemia, and
in the mines of Deux-ponts in the cavities of a
ferruginous clay, mingled with malachite and grey
copper, &c.

COMBUSTIBLE MINERALS.

Including non-metallic substances, the greater part
of which are eminently combustible, and whose
bases are carbon and sulphur.

SULPHUR.

The nature and properties of Sulphur have already been noticed in treating of combustibles generally.

Sulphur is found in the mineral, vegetable and animal kingdoms; in the two latter it occurs so rarely, that all the vast commercial demands for it are supplied from the former source. It is found nearly pure; and is then termed Native Sulphur. It is also found in combination with several of the metals, forming the various pyrites, and the sulphuretted ores. In the state of an acid, it occurs combined with some of the earths and metals.

Native Sulphur is of a pale greenish yellow colour. It occurs in mass, disseminated, in rounded fragments, stalactitic, and crystallized. Its specific gravity is about 2.

It is sometimes, though rarely, found in veins in primitive mountains; its common repository is in beds of secondary gypsum, where it occurs in rounded masses; it is sometimes met with in beds of indurated marl, and compact limestone : occasionally it occurs as an ingredient in mineral waters. Volcanoes abound with sulphur, which sublimes in the rifts and cavities of the lava in the neighbourhood of their craters.

Humboldt mentions its occurrence in a bed of quartz, traversing a primitive mountain of micaceous schistus, in Quito. He also cites two deposites in primitive porphyry.

Sulphur occurs in rounded masses in blue marl in the Apennines of Piedmont. In some of the glaciers of Mont Blanc, it is disseminated in masses of sulphate of lime and clay. At Conilla, near Cadiz in Spain, it occurs in swine-stone. It is met with in the gypsum of the salt springs of Lorraine. It also occurs in Hanover, Hungary, Poland, Siberia, and other countries.

The Warm springs of Aix la Chapelle, of Tivoli, &c. deposit Sulphur when in contact with the air : it is also contained in the waters of certain springs in France.

Volcanic Sulphur is met with in Italy, Iceland, and in Guadaloupe in a volcanic mountain yet in activity. The volcanoes of the Cordilleras in Quito, yield it in great abundance and very pure.

But perhaps the most remarkable deposite of volcanic sulphur is that of Solfatara near Naples, in a kind of sunken plain surrounded by rocks, which is regarded as the crater of an ancient volcano; and from it, ever since the age of Pliny, has been obtained a considerable proportion of the sulphur used in Europe.

The crystals of sulphur are not always well defined : those from Sicily are the best, being frequently perfect ; they have been met with 5 inches in length. The primitive crystal is a very acute octohedron, on which are occasionally found the planes of several modifications. The crystals are often semi-transparent : they are soft, brittle and easily broken.

DIAMOND.

The Diamond, which is the hardest substance in nature, was heretofore considered as an earthy or stony substance; but it is proved beyond a doubt not to be an earthy substance. When exposed to a current of air, and heated to the temperature of melting copper, it is found to be gradually, but completely combustible. By this process it may be wholly converted into carbonic acid, and therefore consists of *pure carbon.*

Diamonds are either colourless, or of a yellowish, bluish, yellowish green, clove brown, black brown, prussian blue, or rose red colour. They are always found in detached crystals, the primitive form of which is the regular octohedron; but the varieties of form in which they occur are numerous. Although the Diamond is so extremely hard, it may be readily cleaved in particular directions. When heated, it becomes phosphorescent. It possesses only a simple refraction, but this may be attributed greatly to its density, considered as a stone. Newton, in remarking this, suspected that the Diamond ought to be placed among combustibles. It is about $3\frac{1}{2}$ times heavier than water.

In India, the diamond mines extend through a long tract of country, from Bengal to Cape Comorin, at the foot of a chain of mountains 50 miles in length: the chief of them are now between Golconda and Masulipatam. Diamonds are also procured from the Isle of Borneo and from Brazil;

where, as well as in India, they are found in beds of ferruginous sand or gravel.

Fifty years ago there were more than 20 places in the kingdom of Golconda in which diamonds of different sizes were found. At that period, 50 workings were also wrought in the kingdom of Visapour. These mines furnished more diamonds than the others; but being smaller, the workings were abandoned. The Diamonds of Pastael, 20 miles from Golconda at the foot of the Gate mountains are the most in request. The mines are situated at the conflux of two rivers; they have produced the most noted diamonds, and amongst them that which has obtained the name of the Pitt or Regent Diamond, the finest of the crown jewels of France, weighing 136 carats, or nearly one ounce, and which was purchased for 2,500,000 livres.

From Mawe's Travels in the Interior of Brazil, we find that the Diamond mines of that country are situated nearly due north of the mouth of the Rio Janeiro. The capital of the district is called Tjuco. The country is covered in all directions by grit-stone rocks, full of rounded quartzose pebbles. The hills are very numerous, and consist of grit alternating with micaceous schistus, and present an immense number of blocks composed of grit-stone imbedding rounded masses of quartz, giving to the whole the appearance of a pudding-stone. The general level of the country must be considerably elevated; it is very full of streams, which fall into the rivers traversing the lower country in almost every direction. Diamonds have been largely obtained in various

places in this district, and always from the beds of
the streams or rivers; most of which have yielded
them. The principal work is that called Mandanga
on the river Jigitonhonha: which being shallow,
though broad, its waters are either dammed out, or
diverted from their course, or pumped out by a
particular contrivance. The mud of the river is
then removed, discovering a stratum of *cascalhao*,
which consists of rounded pebbles and gravel; this
is taken up, and the diamonds are washed out of it.
Diamonds are by no means peculiar to the beds of
rivers or ravines; they have been found in cavities
and water courses, on the summits of the most lofty
mountains of the district.

A diamond, found about 15 years ago in a rivulet
called Abaité, a few leagues north of the Rio Plata,
and now in the possession of the Prince Regent of
Portugal, weighs seven-eights of an ounce. It is
of an octohedral form.

One of the largest known diamonds was in the
possession of the late Empress of Russia; it was of
the size of a pigeon's egg and weighed 193 carats
or nearly one ounce and one-third of an ounce.

The largest diamond hitherto found, is in the possession of the Rajah of Mattan, in the island of
Borneo, in which island it was found about 80 years
ago. It is shaped like an egg, with an indented
hollow near the smaller end. It is said to be of
the finest water. It weighs 367 carats. Now as
156 carats are equal to 1 oz. Troy, it is obvious
that this diamond weighs 2 oz. 169.87 gr. Troy.
Many years ago the governor of Batavia tried to

purchase this diamond. He sent a Mr. Stuvart to the Rajah, who offered 150,000 dollars, two large war brigs with their guns and ammunition, together with a certain number of great guns, and a quantity of powder and shot. The Rajah, however, refused to deprive his family of so valuable an hereditary possession, to which the Malays attach the miraculous power of curing all kinds of diseases, by means of the water in which it is dipped, and with which they imagine that the fortune of the family is connected.

The principal use of the diamond is in ornamental jewellery; it is also employed by glaziers to cut glass, and by lapidaries to engrave the harder gems; but for these purposes such only are used as cannot be cleaved in particular directions.

MINERAL CARBON.

Mineral Carbon is of a greyish black colour, and is destitute of bitumen: it consists of charcoal, with various proportions of earth and iron.

It has a glimmering, silky lustre, and a fibrous appearance, discovering a wood-like texture. It is somewhat heavier than common charcoal, and is easily reduced to ashes before the blowpipe, without either flame or smoke.

It occurs in thin layers in brown coal, slate coal, slaty glance coal, and pitch coal; but in quantities too small to be made separate use of.

Plumbago is found in England, Scotland, France, Spain, Germany, and some other countries. Plumbago is of a dark iron black, passing into steel grey.

It occurs in mass, in kidney-shaped lumps, or disseminated, in rocks. It has a glistening metallic lustre, its fracture is granular and uneven; it is unctuous to the feel, soft, and not very brittle. When heated it does not flame, nor can it support combustion by itself. Its specific gravity somewhat exceeds 2.

Plumbago seems to belong exclusively to primitive countries; sometimes it enters into the composition of rocks: but is more usually found in detached masses, or in beds.

The principal use of plumbago is in the making of what are called *black-lead* pencils; for which purpose none has yet been discovered equal to that from Borrowdale in Cumberland, where it occurs in a considerable mountain of argillaceous schistus, traversed by veins of quartz ; some account of the mine may be found in Parkes's ' Chemical Essays.' An inferior kind is met with in several places in France. It is also found in Bavaria, in Spain, and in Norway.

Whence this mineral obtained the name of *black-lead* it is difficult to say, unless it was from the lead-coloured streak which it gives upon paper. It has been ascertained that lead does not enter into

its composition, but that the purest plumbago consists of about 90 parts of carbon and 10 of iron: an
impure variety affords more of silex and alumine
than of carbon or iron.

MINERAL OIL.

Under this term are comprehended two substances, Naptha and Petroleum; both of which are
liquid, highly inflammable, and lighter than water.

Naptha is nearly colourless and transparent; it
burns with a blue flame, much smoke, gives out a
penetrating odour, and leaves no residuum. It appears to be the only fluid in which oxygen does not
exist in a considerable proportion; advantage has
been taken of this circumstance by Sir H. Davy,
who employed it, for that reason, in preserving the
new metals discovered by him.

The most copious springs of naptha are on the
coast of the Caspian sea in the peninsula of Apche-
ron; the surrounding country is calcareous, and the
soil which affords the naptha is sandy and marly.
It perpetually gives out vapours of a penetrating
odour and very inflammable: it is said that the
people of the country dress their food by means of
it, for which purpose they pass it through earthen
pipes. By distillation it yields naptha pure for
medicine. The Persians employ the residuum to
burn in their lamps instead of oil. A considerable
revenue is derived from it by the Chief of the
country.

Naptha is also found in Calabria; on Mount Zibio near Modena; in Sicily, and in America, &c.; but it is supposed that travellers have sometimes mistaken petroleum for naptha.

In 1802 near the village of Amiano, in the state of Parma, a spring of naptha of a topaz yellow colour, was discovered, which readily burns without leaving any residue; it rises in sufficient quantity to light up the city of Genoa, for which purpose it is employed.

Petroleum, at the usual temperature, is rather thicker than common tar, has a strong disagreeable odour; and is of a blackish or redish brown colour. It is very combustible, giving out during combustion a very thick black smoke, and leaving very little residue in the form of a black coal.

It is found in many countries, principally in those producing coal. At several places in France. In England, at Ormskirk in Lancashire, and at Coalbrookdale; occasionally in Cornwall and in Scotland. It occurs also in Bavaria, Switzerland, and in Italy near Parma. Near the latter place, the Petroleum gives out so powerful an odour, that the workmen cannot long endure it at the bottom of the Petroleum wells, without danger of fainting. It is found in many other parts of Europe and in America.

It is most plentifully found in Asia: round the town of Rainanghong in the Birman empire, there are 520 wells in full activity, into which petroleum flows from over coal. No water ever penetrates into these wells. The quantity of petroleum annually produced by them amounts to more than 400,000 hogsheads. To the inhabitants, its uses are important; from Moussoul to Bagdad it is used instead of oil for lamps ; mixed with earth or ashes, it serves for fuel.

When naptha is exposed to the air and light, it becomes brown, thickens, and seems to pass into petroleum : and when petroleum is distilled, an oil is obtained from it similar to naptha. When petroleum is exposed to the air, it thickens and passes into a kind of bitumen. Considerable alliance is thus proved to exist between Mineral Oil and Bitumen.

BITUMEN.

The elementary constituents of Bitumen are carbon and hydrogen, occasionally nitrogen, and most probably some oxygen, which it is supposed, by its action on the other principles, and in proportion to its quantity, tends to form the harder Bitumens.

Bitumen is either *elastic* or *compact.*

Elastic bitumen is of various shades of brown. It has a slightly bituminous odour, and is about the weight of water. It burns readily with a large flame and much smoke, but melts by a gentle heat, and is thereby converted into a substance resembling petroleum, or maltha, or asphalt, according to its previous consistence.

Elastic bitumen takes up the traces of a pencil in the same manner as the Caoutchouc or India rubber, whence it has obtained the name of *Mineral Caoutchouc.*

Hitherto it has only been found in the Odin mine, near Castleton in Derbyshire, in a secondary limestone, accompanied by calcareous spar, fluor, blende, galena, pyrites, and asphalt. Elastic bitumen consists chiefly of bituminous oil, hydrogen gas and charcoal; very small proportions of other substances have been detected by analysis.

Compact bitumen is of a brownish-black colour; one variety may be impressed by the nail, and is called maltha; another is very brittle, and is called asphalt.

Maltha is brownish black and opake : it is tough, and soft enough to take an impression of the nail : it has a strong disagreeable odour, and is nearly twice as heavy as water. It consists of bitumen mixed with about 8 per cent. of carbon and a little earth.

Maltha is found in France, at a place called Puy de la Pège; where it renders the soil so viscous, that it adheres strongly to the foot of the traveller. It is also found in a mountain in Persia, between Schiraz and Bender-congo, where it is called baume-momie. It is collected with care, and sent to the King of Persia as being efficacious in the cure of wounds. It is occasionally used as a pitch, and in certain varnishes to preserve iron from rust; it is said to enter into the composition of black sealing wax.

Asphalt is brownish black : it occurs in mass, or disseminated, or stalactitic; it is opake, smooth and brittle, and somewhat unctuous to the touch, and gives out when rubbed a slightly bituminous odour. It is not so heavy as maltha.

By combustion, it leaves a small quantity of ashes. It consists chiefly of bituminous oil, hydrogen gas and charcoal, but the latter is in much greater proportion than in elastic bitumen ; oxide of iron, and two or three of the earths, sometimes constitute very small proportions of it.

It is found in the Palatinate; in France; at Neufchatel in Switzerland; in large strata at Aolona in Albania; and in large pieces on the shores, or floating on the surface, of the Asphaltic lake in Judea, called the Dead Sea; which is said to have obtained the latter name from the belief that the Asphaltum caused the death of birds attempting to fly over it. It abounds in the islands of Barbadoes and Trinidad in the West Indies. In the latter it occurs in a vast stratum, three miles in circumference, called the Tar-lake; the thickness of which is unknown. A gentle heat renders it ductile, and when mixed with grease or common pitch, it is used for paying the bottoms of ships, and is supposed to protect them from that pest of the West Indian seas, the teredo or borer.

Asphalt is also employed as a varnish, and an essential part of the best wax, or varnish, for the use of engravers.

Both varieties of Bitumen, as well as the Mineral Oils, are unknown in primitive rocks, except in veins; they seem to belong exclusively to alluvial or primitive formations, in which they most commonly occur in calcareous, or clayey soils, or in the productions of volcanoes. They are said to be mostly found in the neighbourhood of salt forma-

o

tions. Some have conceived that the bitumens and mineral oils have originated in the destruction of the multitude of animals and vegetables found in the earth ; of which we are every day discovering the remains.

The ancients employed bitumen in the construction of their buildings ; and it is said that all historians agree that the bricks of which the walls of Babylon were built, were cemented with hot bitumen ; which gave them very great solidity. Bitumen was carried down by the waters of a river which joined the Euphrates ; it was also found in the salt springs in the neighbourhood of Babylon. The Egyptians are also said to have employed it for the embalming of bodies ; constituting what now we call mummies.

Bitumen enters into the composition of the black indurated marl or shale which accompanies common coal ; and which is generally mixed with it in variable proportions. It is likewise found in certain limestones ; for instance that of Aberthaw, of which bitumen forms about 2 per cent.

COAL.

The bituminous substance called coal, though ranked among minerals because its basis is pure carbon, is now by many believed to be of vegetable origin, because the substance which lies upon the coal, is always filled with vegetable remains ; as well as because a wood-like appearance may be traced through every species of coal, even the most compact.

Mineralogists are not agreed in their arrangement of this important genus of mineral imflammables.

Coal may be divided into four species: brown coal, black coal, cannel coal, and glance coal.

BROWN COAL is imperfectly bituminous, of a brown colour, and of a vegetable texture. It may be divided into three varieties: Bituminized wood, Earthy Brown Coal, Compact Brown Coal, and Moor Coal.

Bituminized wood is of a dark brown colour. Its external shape exactly resembles that of compressed trunks and branches of trees; its internal texture is precisely that of wood, retaining not unfrequently that of the bark. It is opake, soft, somewhat flexible, and almost light enough to float upon water. It burns with a clear flame, though with but little heat, and gives out a bituminous odour, often mixed with that of sulphur.

The surturbrand of Iceland contains 58 per cent. of watery and volatile inflammable matter, leaving 42 per cent. of carbonaceous and earthy residue.

Bituminized wood occurs in alluvial land among beds of compact brown coal; sometimes also forming beds of itself. It is also met with in dispersed fragments in alluvial soil. It abounds in the newest flœtz-trap formation, and forms masses in limestone and sandstone belonging to the independent coal formation. In the Prussian amber mines, it forms the stratum immediately above the amber, and nodules and stalactites of this beautiful substance are generally found intermixed with it.

Earthy Brown Coal is blackish or wood brown, or yellowish grey; it occurs in mass, of a consistence between solid and friable; it is without lustre; soils the fingers a little; it is very soft, falls easily to pieces, and it is a little heavier than water.

It readily takes fire, and burns with a weak flame and disagreeable bituminous odour. It contains 15 to 20 per cent. of earth and oxide of iron, the remainder being water and inflammable matter. It often contains pyrites, and then passes into alum earth.

It is found in similar situations with bituminized wood: in the neighbourhood of Leipsig it occurs in beds from 20 to 40 feet thick, and of great extent.

It is used as an inferior kind of fuel, when little heat is required; for which purpose it is moistened with water, well beaten, and made into masses like bricks. In the vicinity of Cologne, a variety is found of a rich redish brown colour, which is prepared as a pigment under the name of *Cologne umber*, which is employed as a colour both in distemper and oil painting. It is also found in Hesse, Bohemia, Saxony, and Iceland. The Dutch are said to employ it in the adulteration of snuff, to which it imparts, when used in a certain proportion, a peculiar softness.

Compact Brown Coal, is of a blackish brown colour. It occurs in mass; its fracture is fibrous lamellar; its cross fracture somewhat conchoidal; it has a resinous lustre, and is moderately hard. Its specific gravity is about 4.5.

It burns readily with a weak flame, and a disagreeable odour; by combustion, it leaves a small quantity of white ashes : 200 grains yield by distillation, 90 of charcoal, 60 of acidulous water, 21 of thick, brown, oily bitumen, and 29 of hydrogen, carburetted hydrogen, and carbonic acid.

In England, it is found at Bovey near Exeter, and is called Bovey coal : at this place there are 17 strata within the depth of 74 feet from the surface, alternating with alluvial clay; the greatest thickness of the seams or beds is between 6 and 8 feet. Brown coal is also found in various parts of the territory of Hesse, and in other parts of Germany; also in Denmark, Greenland, and Italy.

It is used for fuel : it passes into bituminous wood and moor coal, and sometimes into pitch coal.

Moor coal is of a darkish brown colour. It occurs in mass forming very thick beds, and is always full of crevices. Internally it displays a considerably resinous lustre. Its longitudinal fracture is somewhat slaty; its cross fracture approaches to conchoidal; its fragments are trapezoidal or rhomboidal. In its chemical characters it resembles the preceding variety of Brown coal.

It occurs in alluvial land, and in the newest flœtz trap formation. It is met with very frequently in Bohemia; it is also found in Transylvania, and other parts of the Austrian dominions; also in Denmark, and the Ferroe islands.

BLACK COAL, the SLATE COAL of Mineralogists, is perfectly bituminous ; it may be said to comprehend all the varieties of common coal used for economical purposes. It may generally be said to be of a black colour, having an iridescent tarnish, and a high resinous lustre. It is composed of about 60 parts of carbon, and 36 of maltha and asphalt, and from 3 to 6 per cent. of earth and oxide of iron.

Slate coal is found principally in the independent coal formation, and is the most widely diffused of any of the species. It is often mixed with pyrites, and penetrated by thin veins of quartz or calcareous spar. It always occurs in nearly horizontal strata, which are abundant in Durham, Northumberland, Yorkshire, and in some other parts of England, and in several parts of Europe.

The two points which are principally to be attended to with regard to common coal, in an economical point of view, are the intensity of heat and the duration of combustion, and these are chiefly influenced by the proportion of asphalt contained in the coal. That in which the bituminous part is chiefly maltha, with only a small quantity of asphalt, kindles very easily, burns briskly and quickly with a bright blaze, cakes but little, requires no stirring, and by a single combustion is reduced to loose ashes ; such are the varieties of coal from Lancashire, Scotland, and most of those which are raised on the western side of England.

Those on the other hand, in which asphalt prevails, kindle difficultly, and after lying some time on the fire, become soft, and almost in a state of semifusion ; they then cohere and cake, swell considerably, and throw out on every side tubercular scoriæ, accompanied by bright jets of flame. In consequence of the cohesion and tumefaction of the coals, the passage of the air is interrupted, the fire burns hollow, and would be extinguished if the top were not broken in from time to time. The pro‧ duce of ashes is smaller than in the free burning coals ; the greater part of them being mixed in the carbonaceous part of the coal and forming grey scoriæ, commonly known by the name of cinders, which being burnt again with fresh fuel, give out an intense heat, and are slowly reduced, partly to heavy ashes, partly to slag. The best coal of Northumberland, Durham, and Yorkshire is of this kind ; it burns slower, and gives out more heat than the preceding, and in general bears a higher price.

Cannel Coal or *Candle Coal*, is of a greyish black colour, and has a glistening resinous lustre. Its fracture is conchoidal. It is brittle, but is the most difficultly frangible of all the coals, and is somewhat heavier than jet. It is very inflammable, and burns quickly, but does not cake, and leaves behind 3 or 4 per cent. of ashes.

It occurs in the independent coal formation. It is found in great plenty and remarkably pure, at Wigan in Lancashire, and occasionally in most other English collieries.

Its chief use is as fuel, but the purest from Wigan

may be worked in the turning lathe, from which it receives a high polish; hence it is shaped into various ornamental utensils; and when cut into beads, is not to be distinguished from jet.

The *Splent* Coal, which abounds at Glimerton, near Edinburgh, is considered to be an inferior variety of cannel coal.

Glance Coal. Of this there are three varieties, the conchoidal, the columnar, and the slaty.

Conchoidal Glance Coal is iron black; it occurs in mass, with a bright shining metallic lustre and a perfectly conchoidal fracture. It is moderately hard, frangible, and light.

It burns without flame or smell, and leaves a white ash. It is of rare occurrence, having been met with only at Newcastle and in the Meissner at Hesse. That from the latter place has been analyzed, and contains nearly 97 of charcoal, 2 of alumine, and 1 of oxide of iron.

Columnar Glance Coal is of a dark greyish black; it occurs in mass and possesses a shining lustre, between resinous and metallic. Its fracture is not perfectly conchoidal; is very soft, frangible, and light, and always occurs in thick, curved, parallel, columnar, distinct concretions, having smooth glimmering surfaces.

It burns without flame or smell, leaving a greyish white ash. It has hitherto only been found at the Meissner in Hesse, where it occurs, together with other coal, in the newest flœtz-trap formation.

Slaty Glance Coal, Anthracite, Kilkenny Coal, or *Welch Culm,* is a dark iron black colour, verging on steel grey; it occurs in mass; has a bright metallic lustre; its fracture is somewhat slaty and curved in one direction, and somewhat conchoidal in the other; it breaks easily, and is but little heavier than water.

When pulverized and heated, it becomes red and slowly consumes with a very light lambent flame, without smoke, and when pure emits no sulphureous or bituminous odour; it leaves a variable proportion of redish ashes.

The Kilkenny coal is somewhat harder than is customary with this variety.

Slaty Glance coal consists of carbon, with from 3 to 30 per cent. of earth and iron.

This mineral occurs in imbedded masses, beds, or veins, in primitive, transition, and flœtz rocks. It is found in gneiss, in micaceous schistus, in mineral veins, with calcareous spar, native silver, mineral pitch, and red iron ore; and has been discovered by Jameson in the independent coal formation in the isle of Arran.

JET. PITCH COAL.

Jet is generally of a velvet black; it occurs in mass, and sometimes in the shapes of branches, with a regular woody structure. It has a brilliant, resinous lustre, and a perfectly conchoidal fracture; it is soft and brittle, and is but little heavier than water.

It burns with a greenish flame and a strong bituminous smell, leaving a yellowish ash. It occurs principally in marly, schistose, calcareous, or sandy beds.

It is met with in several places in France : where it is sometimes found enclosing amber. In one place it occurs in oblique beds, at a considerable depth, between beds of sandstone. It is likewise found near Wettemberg in Saxony, and in several places in Spain. It occurs in the Prussian amber mines in detached fragments, and is there called black amber.

In France, Germany, and Spain, it is worked into various trinkets chiefly worn as part of the mourning habit; but when not sufficiently fine and hard for that purpose, it is used as fuel.

AMBER

Is a mineral of a yellow or redish brown, or of a greenish or yellowish white colour. It is found in nodules or rounded masses, from the size of coarse sand to that of a man's head.

It is sometimes transparent, always translucent, and occasionally encloses insects of the ant species, in remarkable preservation. It is somewhat heavier than water. The strong electric powers of amber are generally known. This property gave rise to the science of electricity, which was so called from Ηλεκτρον (Electron) the Greek name for amber. It seems to belong exclusively to countries of late formation.

In Greenland, Kamschatka, and Moravia, it is found in grains disseminated through coal. It also occurs on the shores of the Baltic, of Sicily, and of the Adriatic sea; in Poland, France, Italy, and many other countries; and occasionally in the beds of gravel in the neighbourhood of London, and on the coast of Norfolk and of Suffolk. Near the sea coast in Prussia, there are regular mines of amber: under a stratum of sand and clay about 20 feet thick, succeeds a stratum of trees 40 or 50 feet thick, half decomposed, impregnated with pyrites and bitumen, and of a blackish brown colour. Parts of these trees are impregnated with amber, which sometimes is found in stalactites depending from them. Under the stratum of trees were found pyrites, sulphate of iron and coarse sand, in which were rounded masses of amber. The mine is worked to the depth of 100 feet, and from the circumstances in which the amber is found, it seems plain that it originates from vegetable juices.

The real nature and origin of amber are not understood: it is generally considered to be a fossil resin somewhat mineralized. It yields by distillation an acid called the *succinic acid*, (succinum being the Latin for amber) and leaves as the residue an extremely black, shining coal, which is employed as the basis of the finest black varnishes. When exposed to flame in the open air, amber takes fire and burns with a yellowish flame, giving out a dense, pungent, aromatic smoke, and leaving a light, shining, black coal.

MELLITE. HONEYSTONE.

The Mellite is a rare mineral, having hitherto only been found in Thuringia, in the district of Saal, and in Switzerland. It occurs on bituminous wood, and earthy coal, and is generally accompanied by sulphur. In Switzerland it is accompanied by asphaltum.

The Honeystone is softer than amber, is transparent, brittle, and electric, possesses a double refraction, and is found crystallized in the octohedron.

When burnt in the open air, neither smoke nor flame are observable, and it eventually acquires the colour and consistence of chalk. The Mellite is composed of 84 parts of mellitic acid, about 14 of alumine, 2 of silex, and some iron. Its composition differs essentially from that of every other combustible mineral.

RETINASPHALT.

Retinasphalt has been found at Bovey Tracey in Devonshire, adhering to brown coal in the form of irregular opake lumps of a pale brownish yellow colour, with a glistening lustre and imperfect conchoidal fracture. It is very brittle and soft, and somewhat heavier than water. When placed on a hot iron, it melts, smokes, and burns with a bright flame, giving out a fragrant odour; it consists of 55 parts of resin, 42 of asphalt, and 3 of earth.

FOSSIL COPAL.   HIGHGATE RESIN.

Fossil Copal or Highgate Resin was found in considerable quantity in the bed of blue clay of which Highgate Hill near London, in great measure consists. It is in irregular roundish pieces of a light yellowish dirty brown colour, sometimes transparent and with a resinous lustre ; it is brittle, yields easily to the knife, and is but little heavier than water. It gives out a resinous aromatic odour when heated, and melts into a limpid fluid ; when applied to the flame of a candle it takes fire, and before the blow-pipe burns away entirely.

INDEX.

INDEX.

INDEX.

INDEX.

INDEX.

INDEX.

INDEX.

INDEX.

INDEX.

INDEX.

INDEX.

INDEX.

INDEX.

FINIS.

Printed by W. Phillips,
George Yard, Lombard Street; London.

Printed in the United States
By Bookmasters